African Climate Futures

OXFORD STUDIES IN AFRICAN POLITICS AND INTERNATIONAL RELATIONS

General Editors
Nic Cheeseman, Peace Medie, and Ricardo Soares de Oliveira

Oxford Studies in African Politics and International Relations is a series for scholars and students working on African politics and International Relations and related disciplines. Volumes concentrate on contemporary developments in African political science, political economy, and International Relations, such as electoral politics, democratization, decentralization, gender and political representation, the political impact of natural resources, the dynamics and consequences of conflict, comparative political thought, and the nature of the continent's engagement with the East and West. Comparative and mixed methods work is particularly encouraged. Case studies are welcomed but should demonstrate the broader theoretical and empirical implications of the study and its wider relevance to contemporary debates. The focus of the series is on sub-Saharan Africa, although proposals that explain how the region engages with North Africa and other parts of the world are of interest.

African Climate Futures

CARL DEATH

OXFORD
UNIVERSITY PRESS

OXFORD
UNIVERSITY PRESS

Great Clarendon Street, Oxford, OX2 6DP,
United Kingdom

Oxford University Press is a department of the University of Oxford.
It furthers the University's objective of excellence in research, scholarship,
and education by publishing worldwide. Oxford is a registered trade mark of
Oxford University Press in the UK and in certain other countries

Published in the United States of America by Oxford University Press
198 Madison Avenue, New York, NY 10016, United States of America

British Library Cataloguing in Publication Data
Data available

Library of Congress Control Number: 2025934026

ISBN 9780198960744

DOI: 10.1093/9780198960775.001.0001

Printed and bound by
CPI Group (UK) Ltd, Croydon, CR0 4YY

MIX
Paper | Supporting
responsible forestry
FSC® C013604

The manufacturer's authorised representative in the EU for product safety is Oxford University Press España S.A. of El Parque Empresarial San Fernando de Henares, Avenida de Castilla, 2 – 28830 Madrid (www.oup.es/en or product.safety@oup.com). OUP España S.A. also acts as importer into Spain of products made by the manufacturer.

To Rae, who knows the power of a good story.

To Lola, who accompanied me on this research from the very beginning.

To Rio, whose more-than-human world is livelier than most.

Acknowledgements

To my surprise, this book has become—at least in part—an extended reflection on my personal, intellectual, ecological, and political debts. For this reason, I'll keep the acknowledgements brief. Many people have read sections of the text or discussed it with me and have generously offered comments, including Heather Alberro, Michael Albert, Omayma Al-Khaffaf, Daniel Bailey, Elena Barabantseva, Josh Barritt, Lauren Beukes, Amy Brady, Martin Coward, Paula Diehl, Nelly-Helen Ebruka, Jana Fey, Jake Flavell, Ellie Gore, Sophie Harman, Sophia Hatzisavvidou, Kai Heron, Jenn Hobbs, Amanda Källstig, Aliki Koutlou, Tom Lavers, Gediminas Lesutis, Julija Loginovic, Sherilyn MacGregor, Amanda Machin, Florian Malzacher, Oliver Marchart, Sean McDaniel, Manjana Milkoreit, Matipa Mukondiwa, Yacob Mulugetta, Aoileann Ní Mhurchú, Chuks Okereke, Mitterand Okorie, Mat Paterson, Declan Penrose, Perla Polanco Leal, Liz Richardson, Adrienne Roberts, Cecilie Sachs Olsen, Dan Silver, Youba Sokona, Johannes Stripple, Paul Tobin, Sam Toscano, Silke Trommer, James Wan, Charlotte Weatherill, Robbie Watt, Dan Welch, Nicholas Westcott, Marthe Wilhelmsen, Japhy Wilson, and Harald Winkler. I apologize to anyone I have mistakenly omitted, and I bear all responsibility for the final text.

Thanks are due to those who facilitated, listened and responded to presentations of this material at conferences and workshops in Agder, Birmingham, Cambridge, Manchester, and Johannesburg. The Environmental Politics Group in Manchester has been a particularly warm and supportive community while I have been working on this project. Thanks to participants in reading groups where we discussed fiction and politics, particularly the Feminist Sci-Fi and Critical Global Politics reading groups. The 'Political Imaginaries and Social Transformation' workshops organized by Amanda Machin and Cecilie Sachs Olsen, and the 'Planetary Fiction' workshop organized by Nedine Moonsamy and David Shackleton were inspirational. Early versions of this research were published in *Antipode*, *International Feminist Journal of Politics*, *Millennium*, and *New Political Economy*. Thanks to editors and reviewers for their feedback. Several cohorts of students have shaped this book too, including undergraduate classes on African politics and development, postgraduate classes on critical environmental politics, and dissertation students at all levels.

I am very grateful to the enthusiastic support from the Series Editors of Oxford Studies in African Politics and International Relations: Nic Cheeseman, Peace Medie, and Ricardo Soares de Oliveira. They gave much needed encouragement at a pivotal moment. The reviewers of the book were thorough, insightful, and constructive. Rita Abrahamsen continues to provide immeasurable assistance, seen and unseen, as she has since the start. Dominic Byatt and Phoebe Aldridge-Turner at OUP have provided all the editorial support I could ask for.

I am very grateful to Eric Battle and Jonathan Shapiro for granting permission to use some of their wonderful images in this book. Thanks are also due to Emma Loy-Santelli and Regina Brooks at Serendipity Literary Agency for assistance in securing these permissions.

My family are more visible in this book than in anything I have previously written. Their love and support made it imaginable in the first place. No thanks are ever enough. The inadequacy of words to pay tribute to those nonhuman entities and landscapes which shaped and surrounded this endeavour have been brought into focus by trying to write a book like this, but I offer my gratitude nevertheless. This book owes most of all to Aoileann, who was there to discuss it throughout, gave detailed comments on a full draft text, and is the reason why I believe it is worth continuing to build and rebuild pasts, presents, and futures together.

Contents

List of Illustrations

1
Introduction

A story that begins with a rising sun and a bright new day often signals hope for the future. When the story is about climate change, however, the image of a clear dawn and a hot sun can herald danger. African futures are likelier to be hotter and deadlier than in the past. In February 2024, a heatwave in West Africa led to the heat index reaching 50°C for several days. Wasiu Adeniyi Ibrahim, an official at the Nigerian Meteorological Agency, said: 'with every fraction of a degree of global warming, heatwaves like this will become even hotter.'[1] Images of desertification, drought, and the merciless African sun have become dominant and even clichéd features of imagined future climates. In contrast, the multitude of stories about African climate futures explored in this book reveal much more diversity, and many far more surprising images, than are typically encountered in mainstream climate change discourses. Rejecting tropes of Africa as a passive victim of climate change, doomed to become little more than a blasted wasteland as climate catastrophes bite, this book shows how both long-term official low carbon political strategies and Africanfuturist climate fiction envision diverse climate futures replete with opportunities as well as dangers. African climate futures encompass energy transitions and social revolutions, transformed political agency and human subjectivities, and the possibility of repaired relationships with the past, the West, and with more-than-human worlds.

Typically, most climate narratives about African futures start with images of vulnerability. The Gambia's Long-Term Climate-Neutral Development Strategy 2050 begins by noting that 'The Gambia in recent years has experienced increased frequency and intensity of drought, flooding, coastal erosion, windstorms, high temperatures, and intense and erratic rainfalls.'[2] Totlo Tsamaase's short story, 'Eco-humans', begins in Botswana under an oppressive sun: 'Air prices have risen. Again. 650 degrees, the sun wavers in that summer blue.'[3] A scorching sun features in many other stories too: in Nnedi Okorafor's *The Book of Phoenix* the heroine, pictured in Figure 1, immolates New York and

[1] Carrington and Borràs (2024).
[2] Government of The Gambia (2022, p. 18).
[3] Tsamaase (2022, p. 15).

African Climate Futures. Carl Death, Oxford University Press. © Carl Death (2025).
DOI: 10.1093/9780198960775.003.0001

Figure 1 Eric Battle, 'Who fears the reaper'

© 2015 Eric Battle. Previously published in Nnedi Okorafor, *The Book of Phoenix* (London: Hodder, 2016), p. 217.

western civilization, declaring 'I am the sun. Ten thousand degrees Fahrenheit. Ani has pulled me to the earth. To wipe the slate clean.'[4] In this story the sun is a symbol of revenge, a reckoning for the sins of slavery, colonialism, scientific abuses, and climate change. One of the reasons I am fascinated by fiction is that a story that appears to be about African vulnerability can turn into something else. New ways of being in the world can be imagined.

This is a book about the power of stories. Dominant storylines prop up the status quo and structure familiar ways of seeing the world. Dominant visions of 'the world in 2050' and beyond are predominantly modernist, technology-focused, and centre North American or European people and places. With the exception of stereotypical images of famine, drought, desertification, and nameless 'victims', African characters, places, authors, and stories are typically invisible in dominant discourses of climate politics. Few regions of the world are currently, or are predicted to be in the future, as dramatically affected by

[4] Okorafor (2015a, p. 221).

climate change as Africa, yet African landscapes, political actors, and visions of the future have usually been rendered marginal in global politics and culture.[5] Climate mitigation scenarios used by the Intergovernmental Panel on Climate Change (IPCC), overwhelmingly created by modelling teams based in the Global North, typically assume continued African under-development, neglecting equity considerations and the future energy needs required by the Global South to meet developmental goals. Tellingly, one popular 'future history' casually observes that by the twenty-second century the 'human populations of Australia and Africa, of course, were wiped out'.[6] 'Can't you hear the future weeping?' Ben Okri asks, repeatedly and hauntingly, in his poems about climate change.[7]

Yet stories are also a form of political intervention. The next line of Okri's poem is 'Our love must save the world.' Stories can challenge dominant discourses, imagine alternatives, and envision new relationships in climate-changed futures. Maybe stories of survival will keep us going through dark times ahead? As Nnedi Okorafor's leading character, AO, observes in the first line of the novel *Noor*, set in northern Nigeria: 'I would never do this again. But for the moment, I survived. I went on.'[8] Stories can permit us to imagine better futures, to picture the 'bright-spots' on the horizon.[9] Stories can help explore how we might feel in climate-changed futures and perhaps help us to narrate a path through them. Stories can be fuel for political theorizing, and in this book I use Africanfuturist climate fiction to inspire new ways of challenging and enriching theoretical debates in global climate change politics.

At the core of the book is the argument that Africanfuturist climate fiction can fertilize the imagination of radical, reparative, more-than-human climate politics. We all have debts to pay and climate fiction can help us interrogate who this 'we' is and to whom we are indebted, what reparations might entail, and what it might look like to live convivially alongside more-than-human others in heterotopian, climate-changed futures. Long-term climate strategies and debates about energy transitions and green new deals in the global north and south must confront these questions as a matter of urgency. In this book I argue

[5] For IPCC forecasts of the impact of climate change on Africa, see Trisos et al. (2022). For discussions of how African perspectives on future climates have often been marginalized, see Bassey (2012); Eshun (2003); Mitchell and Chaudhury (2020); Myers (2018); Ndlovu-Gatsheni (2014). For African perspectives on future forecasts and scenarios, see Adesida (1994); Elmandjra (1984); Hamann et al. (2020); Pereira et al. (2021).

[6] Oreskes and Conway (2014, p. 33). On the neglect of equity questions in IPCC scenarios, see Kanitkar et al. (2024).

[7] Okri (2023, p. 14).

[8] Okorafor (2021, p. 3).

[9] Bennett et al. (2016); Hamann et al. (2020).

that fictional stories can help us view our own environments and political relations more critically and suggest that if we want to survive to tell new stories in liveable futures then we need a revolutionary overturning of the current world order, based as it is on carboniferous capitalism, climate coloniality, racialized patriarchy and intersectional oppressions. Africanfuturist climate fiction can help challenge dominant discourses about the places, temporalities, ecologies, and politics of global climate change, and in so doing it is a fertile resource for theoretical debates.

By 'theory', I mean both the generalized views we have about how the world works, and academic traditions of reflecting on these views. Political theorizing often works through storytelling—including analogies, myths, thought-experiments, and illustrative examples—and these stories have constitutive power because how we make sense of the world (re)structures and (re)produces that world. For these reasons, the question of who is able and permitted to imagine and narrate stories about the future is politically crucial. Donna Haraway encapsulates this with a characteristically incisive injunction: '[i]t matters what stories make worlds, what worlds make stories'.[10] In a world structured by slavery and genocide, colonialism and racial capitalism, stories from the Global South have often been erased; this too is a form of constitutive power. In response, this book endeavours to draw attention to, to amplify and re-centre, the contributions of African politicians, intellectuals, and artists to imagined futures. African artistic expression—whether music, sculpture, poetry, fiction, film, or in other forms—has a rich political history of making radical, anti-colonial interventions.[11] The research on which this book rests began with the hope and was nurtured by a profusion of examples that African story-telling traditions might have powerful contributions to reimagining global climate futures.

This book focuses on how climate-changed futures are imagined in Africa and by Africans, and how these future visions can be drawn upon to shape political debates and representations of struggles in the present. There are several interesting puzzles bound up in this. First, the future is inherently unknowable and unpredictable, and yet the science of climate change rests upon predicting the consequences of the current reliance on fossil fuel extraction and combustion. Second, it is difficult to imagine the socio-economic and political landscape of 2050 or beyond, and even harder to convince others that

[10] Haraway (2016, p. 12).

[11] A useful introduction to links between popular culture and radical politics in Africa is provided by Musila (2022). See also Adejunmobi (2016); Branch and Mampilly (2015); Hoad (2007); Krishnan (2014); Obadare (2010).

any one particular representation of the future is plausible or feasible, but persuasive and coherent visions and strategies are necessary for the planning of the socio-economic transitions needed to address climate change. Third, dominant visions of the future reflect the interests and values of the powerful, and yet those interests and values have driven global inequalities and insecurities to the point of climate catastrophe. African and other non-Western visions of the future have tended to be marginalized, even as the peoples of the Global South are rendered some of the most insecure in the face of future climates. *African Climate Futures* is a response to these puzzles. It endeavours to unpack the politics of climate futures and look for visions of the future that might challenge rather than reproduce some of the worst imbalances of contemporary global power relations.

The methodological approach at the core of this book is to study official political strategies and fictional texts side-by-side, as reflections of 'socio-climatic' or 'ecopolitical imaginaries'.[12] Imaginaries are socially shared and collectively held world-views, in this case about how environmental changes will shape and be shaped by socio-political changes. These imaginaries have power in that they draw our attention to some issues, position certain actors, relationships, and places as important, and emphasize particular values and attitudes. I read both long-term climate strategies, and climate fiction novels, short stories and films, as different modes of wrestling with similar questions: how to imagine low-carbon, climate-changed futures and how to explain how 'we' get 'there'.

After the Paris Agreement in 2015, all parties to the UN Framework Convention on Climate Change (UNFCCC) were urged to set out policies which could narrate 'robust visions of the low-carbon futures . . . [and] pathways to visualize their transition to low-GHG economic development' by 2050.[13] The next chapter of this book explores the visions narrated by the strategies of Nigeria, South Africa, Ethiopia, The Gambia, and Zimbabwe. Although important, I conclude that these visions, like most other national long-term strategies elsewhere in the world, are quite restricted ways of imagining the future. They typically aspire to be realistic, technically focussed, and depoliticized. They carry neither the political weight nor the imaginative capacity to reassure us about the climate-changed futures to come or to repair the harms already done. As I write this, in late 2024, it is clear that the status quo of climate change politics is deeply worrying. 'Soon it would be too hot',

[12] See Milkoreit (2017) and Hatzisavvidou (2024), respectively.
[13] Pathak (2017, pp. 2–3).

as J. G. Ballard prophetically warned in one of the earliest climate change novels, *The Drowned World*.[14] The effects of climate change in the present are becoming rapidly evident, the predictions for the future are catastrophic, and political action to avert these consequences and transition to a just low carbon economy and society is inadequate, all over the world.[15] I argue that part of the reason for the lack of genuinely transformative decarbonization pathways is the impoverishment of dominant socio-climatic and ecopolitical imaginaries. In the rest of the book, I argue that climate fiction novels, short stories and films can provide important counter-narratives about the future.

I explore the climate fiction of 15 African authors, including Nnedi Okorafor, Wanuri Kahiu, Lauren Beukes, Chinelo Onwualu, Doris Lessing, Tlotlo Tsamaase, and others. Through classic techniques of speculative and science fiction, such as estrangement and defamiliarization, they offer us a different view of the world to that narrated by official low-carbon political strategies. Climate fiction is not approached here as a way to communicate climate science more widely or think up new technological 'solutions'. Rather, it is a resource for critical thinking; it can help us see the world from radically different perspectives.[16] In the words of American novelist Ursula Le Guin, speculative fiction can, at its best, 'enlarge the field of social possibility and moral understanding'.[17] For Patricia Melzer it can provide a 'blueprint' for critical theorizing, a way to 'develop the terminology to describe our future'.[18] I seek to show that Africanfuturist climate stories can play a politically crucial role in foregrounding debates about climate justice, revolution, and restitution, and identity, memory, and trauma. These are debates that are often neglected in mainstream climate change discourse. Africanfuturist climate fiction can challenge how we imagine future cities, families, and relations with other species. These stories insist upon the need to pay attention to the traces of the past in the present and future, and think about how to preserve, memorialize, and repair that which has been bequeathed to the present and the future. The greatest strength of Africanfuturist climate fiction, I suggest, lies in how it can reimagine human subjectivity in future climate-changed contexts, and demonstrate how embracing human difference—racial, gendered, sexual, differently-abled, neuro-diverse—will be vital in coming to terms with changing climates.

[14] Ballard (1962, p. 1).

[15] Big-picture analyses of contemporary environmental and climate politics include Albert (2024); Malm (2021); Newell (2021); Patel and Moore (2017); Paterson (2021a); Sultana (2022); Sze (2020); Yusoff (2018).

[16] On art as a form of critique, as a way of seeing the world differently, see Alberro (2024); Hammond (2021); Haraway (2016); Jackson (2020); Seymour (2013); Streeby (2018); Yusoff and Gabrys (2011).

[17] Le Guin (2016, p. 376).

[18] Melzer (2006, pp. 2–3).

African climate fiction emerges from much older traditions of African self-writing, speculative fiction, and anticolonial Afrofuturist/Africanfuturist imaginaries, as well as the revolutionary and utopian traditions of Black political theory.[19] African stories—in various forms, including oral traditions, novels, short stories, poetry, and films—are particularly important ways through which the past, present, and future are understood and re-examined in African philosophy and political theory.[20] African science fiction is relatively small as a genre, but is growing rapidly and it has considerable political relevance. As Nigerian-American novelist Nnedi Okorafor has argued, 'science fiction is one of the greatest and most effective forms of political writing. It is all about the question: what if?'[21] This question, I suggest, is central to the contemporary politics of climate change.

This sense of urgency and concern for our futures is the primary motivation for writing this book. I have been researching environmental politics in South Africa since the early 2000s, and the effects of climate change—both physical and political—have grown exponentially over this time. Political and economic struggles across the African continent are being and will increasingly be profoundly shaped by climate change, in ways that are likely to worsen the prospects for peace, justice, equality, sustainability, and security. Africa is part of the global economy and climate change is a global concern: Africa's futures will shape, and be shaped by, the futures of everyone on the planet. African climate futures are also intertwined with climate futures in Europe, and hence my personal climate futures, however remote these might seem from my current location in the UK. I am shaped by historical structures that have produced African climate presents and futures: colonialism, industrialization, neoliberalism, globalization, racism, sexism. My everyday consumption patterns—coffee from Ethiopia, chocolate from Ghana, fruit from South Africa, iPhones and iPads assembled from coltan in the Congo, Nigerian oil exports to the EU, novels and films from across the continent and diaspora—are deeply implicated in extractive political economies that shape the unfolding climate catastrophe.[22] Despite the mutual implication of African pasts, presents, and futures with Western political economies, African imaginaries are typically rendered invisible in contemporary climate change politics. Internationally successful published fiction can be a way to challenge that. Lauren Beukes' detective-noir novel *Zoo City* begins with an image of

[19] Key texts here include Gilroy (2005); Jackson (2020); Kelley (2002); Mbembe (2002); Womack (2013); Zamalin (2019).

[20] For examples of this, see Aiyetoro and Olaoye (2016); Mbembe (2001, 2006, 2017, 2019); Nyamnjoh (2017); Opondo (2022).

[21] Okorafor (2017).

[22] Patel and Moore (2017).

the scars of past and present mineral extraction: '[m]orning light the sulphur colour of the mine dumps seeps across Johannesburg's skyline and sears through my window'.[23] Making African visions of the future more prominent and visible is an important political motive for writing this book.

There are three main gaps in academic research that this book seeks to address. First, the extensive and growing literature on the international politics of climate change and environmental issues does not adequately explore how the proliferation of visions of climate futures can shape contemporary politics. While there is considerable research on transitions to low-carbon futures in International Relations and International Political Economy, there is little systematic research on the national plans submitted to the UNFCCC in the form of long-term climate strategies.[24] There is a small but growing body of research on the role of cultural imaginaries and visions of the future in climate politics. Much of this research is focussed on cases from the Global North and African visions of climate futures are almost invisible in this literature.[25] This book is the first sustained study of socio-climatic imaginaries that combines analysis of policy visions and climate fiction in order to assess their strengths and limitations in the broader context of global climate politics, engaging with climate visions from Global South perspectives. It makes the argument for a more central place for political and cultural imaginaries from Africa in the study of energy transitions and low-carbon futures.

Second, within African Studies there is still relatively little critical scholarship on climate change politics or how future climates are imagined. This book is the first major study of African long-term climate strategies and future imaginaries. While there is extensive research on climate change in Africa, and some analysis of long-term climate strategies and energy transitions, this is dominated by single-country studies, especially in South Africa and Nigeria.[26] This book provides the first cross-country study of the handful of long-term strategies submitted before 31 December 2023 by African countries to the

[23] Beukes (2010, p. 3).

[24] On low carbon transitions in IPE, see Hochstetler (2020); Newell (2021); Paterson (2021a); Pearse (2021); Swilling et al. (2016). On long-term strategies, see Death (2024a); Pathak (2017); Ross and Fransen (2017).

[25] On cultural imaginaries of climate change, see Beck (2018); Bina et al. (2020); Hajer and Versteeg (2019); Harris (2017); Hatzisavvidou (2024); Jasanoff (2015); Jessop (2010); Levy and Spicer (2013); Marquardt and Nasiritousi (2022); Paterson (2016); Tozer and Klenk (2018); Vervoort and Gupta (2018); Wood et al. (2024).

[26] Some important exceptions include Cairns et al. (2022); Côté-Roy and Moser (2019); Iheka (2017, 2021); Medhin and Mekonnen (2019); Müller et al. (2020); Myers (2018); Okereke et al. (2019); Simmet (2018). Research on South Africa and Nigeria includes Hochstetler (2020); Nwozor et al. (2021); Osunmuyiwa et al. (2018); Swilling et al. (2016).

UNFCCC. It is also the first to situate these in the wider context of African socio-climatic imaginaries, including Africanfuturist novels, short stories, and films.

Third, this book seeks to contribute to the rapidly growing body of scholarship on climate fiction in the Environmental Humanities. Climate fiction—sometimes referred to as 'cli-fi' after journalist Dan Bloom coined the term in 2007—refers to fictional treatments of climate change caused by human activity in terms of setting, theme or plot, and is rapidly growing by all accounts.[27] Canonical authors and novels include *The Road* by Cormac McCarthy; Liz Jensen's *The Rapture*; *Solar* by Ian McEwan; *The Broken Earth* trilogy by N. K. Jemisin; the *Science in the Capital* trilogy; *New York 2140* and *Ministry for the Future* by Kim Stanley Robinson; *Flight Behaviour* by Barbara Kingsolver; *The Windup Girl* by Paolo Bacigalupi; the *MaddAddam* trilogy by Margaret Atwood; Saci Lloyd's *The Carbon Diaries 2017*; and many others. So far, both the fiction itself and the academic research on it mainly focuses on European and North American settings and characters. African climate fiction is under-researched even amongst the smaller body of research on stories from the Global South. There is very little research on interactions or comparisons between policy and popular imaginaries, particularly from postcolonial, feminist, and queer perspectives.[28] Yet the body of Africanfuturist climate fiction is growing rapidly, and in this book I argue that the work of authors like Nnedi Okorafor, Suyi Davies Okungbowa, Wanuri Kahiu, Lauren Beukes, Chinelo Onwualu, Tlotlo Tsamaase, and others can provide more diverse and politically radical ways to imagine climate futures than are typically introduced into political debate. As Mitchell and Chaudhury argue, 'imaginaries create worlds, so it matters greatly whose are privileged, and whose are excluded'.[29] This book seeks to emphasize the importance of not just incorporating these Africanfuturist authors into the climate fiction canon, but proposes that reading these stories as a form of critical theory can challenge how we understand the places, temporalities, ecologies, and politics of imagined climate futures.

[27] Barnett (2023); Brady (2019); Harper (2023); Johns-Putra (2015); Pattee (2023); VanderMeer (2023); Yoder (2023).

[28] A sample of climate fiction research which reflects this European and North American focus includes Cole (2022); Johns-Putra (2019); Kaplan (2016); Nikoleris et al. (2019); Schneider-Mayerson (2018); Svoboda (2016); Trexler (2015); Thaler (2022); Yusoff and Gabrys (2011). For discussion of climate fiction from a Global South perspective, see Ghosh (2016); Milner and Burgmann (2018); Mitchell and Chaudhury (2020); Streeby (2018); Whyte (2018).

[29] Mitchell and Chaudhury (2020, p. 5).

Theoretical Approach

This book has theoretical concerns—in the sense of debates about differing ontological and epistemological worldviews—at its centre. It starts from the claim that 'present futures' are powerful and have political consequences.[30] As cultural theorist Kodwo Eshun puts it, power 'functions through the envisioning, management, and delivery of reliable futures'.[31] These future visions can be analysed as forms of socio-climatic imaginaries, defined by Manjana Milkoreit as 'collectively held visions of the future, both desirable and undesirable, that are informed by science and can support deliberation and decision-making in the present'.[32] The closely related concept of ecopolitical imaginaries is used by Sophia Hatzisavvidou to shift attention from psychological and cognitive processes to collective struggles and institutional structures, with more emphasis on the envisioning of 'possibilities for political arrangements—through organization and collaboration, negotiation and contestation—that might lead to more sustainable, just, and desirable futures'.[33]

The concept of the imaginary is related to, but distinct from, the concept of imagination. Imagination is often defined in terms of an individual capacity to 'generate ideas in the mind about things that cannot be perceived with the senses'.[34] Crucially, just because something cannot be perceived with the senses does not mean it is not (potentially) real or significant. Capitalism depends upon imagined futures and fictional expectations around mechanisms such as competition and credit. Everyone uses their imagination to make sense of the past and the present and project the possible consequences of their actions into the future. Imagination is not disconnected from the physical and social worlds in which we are all embedded, rather it 'is situated; our imaginary horizons are affected by the positioning of our gaze'.[35]

One of the risks of the concept of imagination is that it can, misleadingly, be used to imply individual mental genius or inspiration. In contrast, the concept of the imaginary is used to refer to socially shared and, to some degree, institutionally supported or stabilized mental pictures. Imaginaries are how people collectively imagine their social surroundings, 'carried in images,

[30] Adam and Groves (2007, p. 28).
[31] Eshun (2003, p. 289).
[32] Milkoreit (2017, p. 14).
[33] Hatzisavvidou (2024, p. 11).
[34] Moore and Milkoreit (2020, p. 2).
[35] Stoetzler and Yuval-Davis (2002, p. 327). Important contributions to the political theory of imagination and imaginaries include Beckert (2016); Bottici and Challand (2011); Castoriadis (1987); Haiven (2014); Jasanoff (2015); Levitas (2013); Said (1994); Taylor (2002); Tilly (2002); White (2011).

stories, and legends'; or the 'interdependent cognitive and social processes that create representations of present and possible future states of the world'.[36] Whereas Charles Taylor uses social imaginaries to refer to a 'common understanding that makes possible common practices and a widely shared sense of legitimacy', among 'society as a whole', others emphasize the existence of multiple, often conflicting, social imaginaries within and across societies.[37] This emphasis on the diversity and instability of relations between conflicting imaginaries encourages a recognition of the inherently radical capacity of the imagination: it is always possible to picture things otherwise and to envisage alternative social relations. Indeed, the work of critical theorists and eco-critics has often been targeted at 'reimagination', or challenging and shifting dominant imaginaries.[38]

Such ecopolitical imaginaries are crucial elements of a just transition to a low carbon future, especially because an inspiring narrative and vision are often argued to be necessary for convincing audiences about the desirability and feasibility of transition.[39] In the US context, Naomi Klein and Alexandria Ocasio-Cortez have drawn attention to the 'incalculable power of art' for achieving a Green New Deal, invoking the example of President Franklin D. Roosevelt's New Deal programme in the 1930s, where the imagination and visions of artists and authors like Zora Neale Hurston, Ralph Ellison, and John Steinbeck played an important role in 'overcoming the scepticism that humanity could ever pull off something at this scale and speed'.[40] For Matthew Paterson, 'an imaginary is a powerful part of the process of creating a set of self-sustaining processes of social transformation'.[41] Ecopolitical imaginaries can encompass both policy texts and fictional stories, as even an individually authored story becomes a wider cultural text once it is edited, published, commodified, distributed, discussed, reviewed, critiqued, and taught. Published fiction is also socially shared and institutionally supported, albeit in different ways than official political strategies and visions.

My analysis of socio-climatic and ecopolitical imaginaries is situated within a broader theoretical orientation inspired by postcolonial, feminist, and queer perspectives. First, my view of the world is shaped by postcolonial theory,

[36] See Taylor (2002, p. 106) and Moore and Milkoreit (2020, p. 12), respectively.

[37] Taylor (2002, p. 106). In contrast, see Jasanoff (2015); Machin (2022); Marquardt and Nasiritousi (2022); Wright et al. (2013).

[38] Clark (2015, p. 19). See also Bottici and Challand (2011); Haiven and Khasnabish (2014); Imarisha and Brown (2015); Kelley (2002); Patel and Moore (2017).

[39] Hatzisavvidou (2024). See also Alberro (2024); Jasanoff (2015); Levy and Spicer (2013); Milkoreit (2017); Newell (2021, p. 203); Ross and Fransen (2017); Vervoort et al. (2015).

[40] Klein (2019).

[41] Paterson (2016, p. 18).

which emphasizes how global politics is 'both historically constituted through colonialism and presently delineated by struggles over colonial legacies'.[42] These legacies are intensified by processes of climate change, as vividly expressed by Farhana Sultana, who charts how climate coloniality constitutes an ongoing wound 'that governs and structures our lives'.[43] The hierarchies, binaries, and Othering through which coloniality structures global politics are always contested and resisted, including through what Sultana describes as 'a multiplicity of local voices, stories, ideas, cosmologies, strengths, and convivialities'.[44] I draw on the political theory of Achille Mbembe—as does Sultana—in order to express how global politics is constituted through formations of biopolitics and necropolitics, or the governing of bodies through practices of life and death. Mbembe's dissection of how structures of power are enacted and resisted through discourses, myths, and stories, as well as his normative commitment to searching for better ways to live—more safely, more ethically, more justly—in an era of more-than-human planetary entanglements, is an inspiration for the analysis in this book. Similar concerns with the ontologies, epistemologies, and ethics of relating to othered humans as well as non-human others, in a world structured by colonial, imperial, and capitalist violence, are explored with insight by theorists like Amitav Ghosh, Cajetan Iheka, and Zakiyyah Iman Jackson. Their work has guided my approach to studying literary texts as a form of postcolonial political theory.[45] Exploring different sets of fictional narratives, these theorists have cast illumination on ways to restore 'agency and voice to nonhumans'; become more 'open toward the Other ... in compassion and empathy'; and provoke 'unruly yet generative conceptions of being'.[46]

Feminist theory is the second theoretical tradition that has inspired this research. Complementing and challenging postcolonial perspectives on legacies of empire, feminists have insisted upon the presence of the political in the personal and the intimate, as well as analysing the reproduction of and resistance to dominant relations of gender and power in language, categories of thought and practices of social reproduction. The work of Donna Haraway has been particularly influential in terms of the role of speculative fiction in challenging dominant figurations of gender, race, class, and species, and the

[42] Rutazibwa and Shilliam (2018, p. 1). See also Sabaratnam and Laffey (2023).
[43] Sultana (2022, p. 3).
[44] Sultana (2022, p. 10).
[45] Ghosh (2016, 2021); Iheka (2017, 2021); Jackson (2020); Mbembe (2001, 2019).
[46] See Ghosh (2021, p. 204); Iheka (2021, p. 161); and Jackson (2020, p. 4) respectively. See also Abrahamsen (2003); Ashcroft (2017); Said (1994); Tamale (2020); Weheliye (2014); Womack (2013); Zamalin (2019).

struggle to imagine new ways of relating to more-than-human others.[47] The potential for fiction to reinscribe and resist dominant intersections of power has been further explored by theorists like Adeline Johns-Putra, Sylvia Tamale, and Patrica Melzer. Johns-Putra is a leading theorist of contemporary climate fiction, including examinations of the representation of parenthood and motherhood in a wide range of novels and reflections on the role literature can play in cultivating an awareness of ethics, empathy, and emotion.[48] Tamale brings feminist and decolonial debates into conversation by observing the differences between the concerns of African feminists to feminists elsewhere in the world, as a result of the way that 'the African continent occupies a separate cultural, social, economic and geopolitical landscape from the West'.[49] She calls for a stronger intersectional Pan-Africanism united by anti-imperialism, anti-patriarchy, and antimilitarism.[50] Melzer takes up Haraway's claim that science fiction is political theory and develops it further, arguing that 'some science fiction texts not only incorporate feminist theory but actually produce it'.[51] She emphasizes 'the spaces of agency located within the process of consuming and producing cultural texts', and insists upon the importance of 'imagination, narrative, and desire as part of feminist theorising'.[52] While I see important differences of form, audience, and (often) style of writing between fictional and academic texts, and want to stress that the theoretical implications we draw from fiction are not pre-existing but must be articulated by the reader/writer, this book is inspired by the feminist lesson that theorizing—like politics—takes place everywhere, including in spaces traditionally defined as domestic, personal, private, and intimate.[53]

The third strand of theoretical literature that has shaped this book is queer theory, in particular the work of authors such as Nicole Seymour, José Esteban Muñoz, and Rahul Rao.[54] This research represents some of the best traditions of reflection and continual critique in critical theory, committed to sustained interrogation of positionality and representation, language and silencing, balancing attentiveness to the constant presence of multiple violences as well as the ineradicability of diverse expressions of hope for better futures. The term 'queer' is used here, following Seymour, as 'that which questions the

[47] Haraway (2016).
[48] Johns-Putra (2016, 2019).
[49] Tamale (2020, p. 40).
[50] Tamale (2020, p. 382).
[51] Melzer (2006, p. 11). See also Haraway (2016).
[52] Melzer (2006, p. 34).
[53] See also Gaard (2017); Grosz (2004); MacGregor (2006); Sandilands (2017).
[54] Seymour (2013, 2018); Muñoz (2009); Rao (2020). See also Butler (2017); Gore (2018); Hoad (2007); Reddy et al. (2018); Otu (2021).

naturalness, and undermines the stability, of established categories of sex, gender, and sexuality'.[55] Many of these theorists wrestle with how queer theory's distrust of concepts like nature and progress can be reconciled with visions of better ecological futures. In the African context, there have been extensive debates over the applicability of terms like queer (paralleling similar debates about feminism), and while there are important political decisions involved in labelling people or groups as queer, same-sex, LGBTQ+, gay, and so on, theorists like Zethu Matebeni and Stella Nyanzi have defended the use of queer theory in African contexts to describe that which 'pushes the boundaries of what is embraced as normative'.[56] It is in this spirit that I have been drawn to authors who use queer theory to argue, like Seymour, that climate change raises challenges about how to relate and to care, politically, for myriad others in ways that might be uncomfortable, ambiguous, irreverent, and which transgress conventional forms of social, familial and financial ties. Muñoz argues that art and culture can assist with thinking through these challenges, casting 'anticipatory illumination' that can 'provide us with access to a world that should be, that could be, and that will be'.[57] In a similar vein, Rao suggests that fictional texts 'can help to interrogate the imaginative limits of hegemonic public cultures'.[58] My initial motivation for turning to queer theory was to help understand the prominence of characters in Africanfuturist stories who identify as queer or gay, as well as same-sex relationships, for example in the novels *It Doesn't Have to Be This Way*; *Lagoon*; *Mara and Dann*; and *War Girls*.[59] More substantively, the rich critical provocations of theorists like Seymour, Muñoz, Rao, Matebeni, and Nyanzi have challenged and pushed forward my analysis of the ways in which these stories narrate place, temporality, and ecological relationships.

These three theoretical approaches each draw attention to 'perspectives and people excluded from prevailing purviews'.[60] As I show in Chapter 2, 'Dominant narratives in long-term strategies', coloniality, race, gender, and sexuality are almost invisible in mainstream long-term low-carbon political strategies. In contrast, Africanfuturist climate fiction makes them central to the places, times, ecologies, and politics of climate-changed futures. Fictional stories can, in this way, contribute to theoretical debates in global politics. In framing these approaches as sets of debates—for example, over what reparations mean

[55] Seymour (2013, p. 28).
[56] Matebeni (2014, p. 7). See also Nyanzi (2014) and Rao (2020, p. 16).
[57] Muñoz (2009, p. 64).
[58] Rao (2014, p. 202).
[59] Mackay (2022); Okorafor (2014); Lessing (1999, 2006); Onyebuchi (2019).
[60] Bleiker (2009, p. 2).

for climate politics, or the applicability of feminist and queer frameworks in African contexts—I also reaffirm the importance of continual critique and reflection. Theorists like Achille Mbembe and Donna Haraway have a landmark status in this field and for this research, but I don't intend to suggest that they are infallible or flawless. Their work has received important criticisms from others, including the neglect or problematic discussion of gender and apparent dismissal of some forms of materialist resistance in Mbembe's work, and some significant tensions and problems in how Haraway discusses race, disability, and multispecies assemblages, for example.[61] Rather than dwelling on the limitations of particular authors or theoretical texts, in this book I have instead chosen to focus on exploring and amplifying imaginaries that have the potential to challenge dominant ways of envisioning the future.

Africanfuturist stories are an excellent place to look for counterhegemonic narratives. The term Africanfuturist was coined by Nigerian-American author Nnedi Okorafor in 2019, situating her work in relation to Afrofuturism. Afrofuturism has been described by South African novelist Lauren Beukes as 'an artistic, aesthetic and philosophical movement that combines science fiction, magic, traditional beliefs, black history and culture.'[62] For Okorafor, in contrast to the African-American roots and US-dominated context of Afrofuturism, 'Africanfuturism is specifically and more directly rooted in African culture, history, mythology and point-of-view as it then branches into the Black Diaspora, and it does not privilege or center the West.'[63] She explains further that Africanfuturism is

> concerned with visions of the future, is interested in technology, leaves the earth, skews optimistic, is centered on and predominantly written by people of African descent (black people) and it is rooted first and foremost in Africa. It's less concerned with 'what could have been' and more concerned with 'what is and can/will be.' It acknowledges, grapples with and carries 'what has been.'[64]

I read Okorafor's work—as well as the other authors and artists discussed in this book—as a form of imaginative politics, a contribution to conversations

[61] For critiques of Mbembe, see Diagne (2002); Vergès (2002); Weheliye (2014). For critiques of Haraway, see Davis et al. (2019); Kafer (2013); Sandilands (2017). Elizabeth Grosz is another theorist whose work has been influential for this project but is also deeply flawed, for example in terms of extremely problematic transphobic commentary. See Seymour (2017, p. 256).

[62] Beukes (2017). On Afrofuturism, see also Brooks et al. (2022); Dery (1994); Eshun (2003); Iheka (2021); Mbembe (2019, p. 163); Pereira et al. (2021); Samatar (2017); Womack (2013); Zamalin (2019).

[63] Okorafor (2019).

[64] Okorafor (2019).

about race, identity, gender, nature, temporality, reparations, and political agency pursued by the postcolonial, feminist, and queer theorists discussed above. These conversations unfold over the course of the book; it cannot be neatly separate into theoretical and empirical chapters but rather seeks to weave more generalized and more localized threads together in an exploration of how Africanfuturist climate fiction can challenge and expand the ways in which different places, times, ecological relations, and politics are imagined.

Global Climate Politics

At this point, some observers of global climate politics might want to ask: why is it important to focus on African long-term climate strategies and imaginaries? African contributions to global greenhouse gases are very small, and global climate politics, one might assume, will be far more profoundly shaped by the imaginaries shared and enacted in the United States, the UK and EU, China, and India. African countries do not need to aim for net zero by 2050 in order for the world to limit warming to 1.5°C. This is all true, but African strategies and imaginaries are crucial in several respects. First, fossil fuel production and consumption are rapidly expanding in Africa; many African economies and cities are growing fast; and by 2050 it is predicted that a quarter of the world's population will be African.[65] Many African countries have huge capacity for renewable energy as well as immensely valuable ecosystems and resources. Second, countries like Nigeria and South Africa are major producers and exporters of oil and coal, respectively, and as early and enthusiastic advocates for energy transitions and low carbon development strategies their lead is likely to be influential for many other fossil fuel producing emerging economies worldwide.[66] Both countries have made political commitments to mid-century net zero targets and South Africa has received $8.5 billion in development finance for their Just Transition, mainly in the form of loans. Nigeria was the first, and by December 2023, the only member of the Organisation of the Petroleum Exporting Countries (OPEC) to have submitted a long-term climate strategy to the UNFCCC. Third, there is an urgent imperative to address historic and ongoing ecological and climate debts owed by the Global North to the Global South. These debts are not only ignored or denied in many climate strategies or iterations of Green New Deals, but typically assumptions about creditors and debtors are reversed such that climate finance is added to the financial burden ostensibly owed by the poor to the

[65] Pereira et al. (2021); The Economist (2020).
[66] Death (2016, 2024a); Hochstetler (2020).

rich.[67] Furthermore the vulnerability of some African societies and economies to future climate impacts, and the marginalization of many African perspectives from global debates about climate politics, means that focusing on and amplifying African imaginaries is crucial from a global justice perspective. As Granjou and Salazar point out, 'too often, ideas about the meaning of "the future" and the possibilities for shaping the "not yet" are not globally shared', and non-western perspectives, in particular, have 'too often been pushed to the fringes'.[68] Considering African-authored visions of the future and demands for reparation is important in terms of climate justice, and there is also a need to encourage democratic debates about climate futures within African countries.

A second important question is: why read climate fiction alongside political strategies? Shouldn't social scientists be focused on the facts, and political interventions that matter in the real world, rather than invented fantasies or speculative futures? Even a leading ecocritic and cultural theorist such as Timothy Clark expresses doubts about the 'viability of ecocriticism itself as a possible force of significant change'.[69] These doubts have been a constant presence whilst doing this research and are worth reflecting on, but I want to argue that fictional visions of the future do have power and do matter for climate politics. They are representations of the world—whether realistic or fantastical—that can reveal a great deal about power relations in the present. They can expand political possibilities and help understand alternative perspectives on class, race, gender, and intersectionality beyond our own experiences and worldviews. Fictional stories can personalize and humanize somewhat abstract issues of climatic changes and energy transitions, and ultimately research in the humanities can help us understand what makes us human and how we relate to non-human others. For Beukes, 'the stories we tell ourselves about ourselves shape who we are—and who we can be'.[70] I am not, in this book, concerned with the direct impact these stories have on those reading them. On this, much research advises caution about assuming that fiction can be relied upon to produce better eco-activists and climate conscious consumers.[71] Instead, I am interested in the imaginative illumination that can be provided to the theorization of climate politics by reading these Africanfuturist stories.

[67] Sabaratnam and Laffey (2023, p. 174).

[68] Granjou and Salazar (2016, p. 242). See also Bassey (2012) and Pereira et al. (2021); as well as, for older perspectives, Adesida (1994) and Elmandjra (1984).

[69] Clark (2015, p. 198).

[70] Beukes (2017). On the power of fiction, see Alberro (2024); Bleiker (2009); Braun (2022); Ghosh (2016, 2021); Milkoreit (2017); Nikoleris et al. (2019); Jackson (2020); Johns-Putra (2019); Streeby (2018); Weldes and Rowley (2015).

[71] For research on this, see Milkoreit (2019); Schneider-Mayerson (2018); Weik van Mossner (2017, p. 160).

In this book I seek to make the case for the importance of studying climate fiction, but it is also important not to romanticize these novels, short stories, and films. Fiction can also reproduce prejudices, mislead, narrow horizons, mask alternatives, and offer a depoliticizing escape from immediate circumstances. Amitav Ghosh has argued that we live in a period best described as '[t]he Great Derangement ... when most forms of art and literature were drawn into the modes of concealment that prevented people from recognising the realities of their plight'.[72] This provocation is important and the ideological function of art is worth examining carefully, as in the work of trail-blazing scholars like Edward Said.[73] Moreover, African art has often been romanticized and exoticized by outsiders seeking something new, exciting and different. African artists and theorists have faced dilemmas over whether to assert universalist humanity or defend nativist particularism.[74] With these pitfalls in mind, I explore whether Africanfuturist climate fiction can play a role in reinforcing or challenging dominant ecopolitical imaginaries. I conclude it has the potential, as Achille Mbembe demands, to help articulate a language that 'knows how to become a projectile' and that can not only 'force the locks' but also 'save life from the disaster lying in wait'.[75]

Methodological Approach and Outline of the Book

There are two main stages to the approach taken in the book. The first stage—and the next chapter of the book, 'Dominant Narratives in Long-term Strategies'—is an assessment of mainstream storylines about climate futures, in the form of African national long-term strategies.[76] Here, I explore the strengths and limitations of the ways in which African political strategies envision the world in 2050. It makes the argument that the wider scholarship on energy transitions would be enriched by closer interrogation of policy narratives and socio-climatic imaginaries, and asks: how do these policy strategies imagine climate futures, and what are the political strengths and limitations of their capacity to envision the world in 2050? I focus on the long-term strategies submitted to the UNFCCC by African countries, specifically the

[72] Ghosh (2016, p. 11).
[73] Said (1994). See also Milkoreit (2016, p. 188); Smith (2012); Weldes (2001).
[74] Mbembe (2002).
[75] Mbembe (2019, p. 189).
[76] These national long-term strategies are not the only form of climate futures: there is a growing literature on models, scenarios, pathways, forecasts, backcasts, etc. from a variety of perspectives. See Albert (2024); Death (2022b); O'Neill et al. (2017); Pereira et al. (2019); Pulver and VanDeveer (2009).

English-language submissions of Nigeria, South Africa, Ethiopia, The Gambia, and Zimbabwe, and conclude by contrasting them with the African Union's *Agenda 2063: The Africa We Want*. Drawing on the literature on narratives in policy texts, I propose that there are four key sets of questions to ask when exploring the ecopolitical imaginaries in these policy visions.[77] First, what is the plot? Where do they start and end? What are the obstacles to be overcome? What theories of change are implied or assumed? Second, who are the main actors in these stories? Are they cast as heroes, victims, or villains? Third, what is the over-arching moral, message or discourse? Fourth, what are the salient silences and gaps in the narrative? These questions are addressed through close textual analysis of the long-term strategies themselves, supported by wider research on energy and climate politics in the countries concerned.

Chapter 2, 'Dominant Narratives in Long-term Strategies', shows that these strategies all narrate national governments in leading roles, driving the transition to low carbon economies. With varying degrees of depth and detail, they all call upon international financial support to play a supporting role. All the strategies begin by emphasizing the vulnerability of the country to the impacts of climate change. They all also emphasize their status as developing countries for whom a growth and poverty-alleviation focus is primary. They all mention the importance of a just transition, although in different ways and with different emphases and scope. Ultimately, I conclude that there are some important shared silences in the ways that they envision what low carbon development entails, how to get to net zero, and the political implications for democracy, subjectivity, and ecology. None of them make a strong case for climate reparations or the climate debt owed by the Global North to the Global South.

The second stage of the research begins in Chapter 3, 'Why Read African Climate Fiction?' This chapter turns to African climate fiction to explore alternative ecopolitical imaginaries in novels, short stories, and films. I treat these Africanfuturist stories more as a corpus of diverse visions of the future from which theoretical insights might be gleaned, than an empirical sample representative of African climate fiction more generally. In my reading I was drawn to fictional stories that were either set in Africa or created by an African author; stories that had prominent climate and/or environmental themes; and stories that imagined an alternative future. I have chosen to focus upon novels by Lauren Beukes, Doris Lessing, Alistair Mackay, Nnedi Okorafor, and Tochi

[77] Beck (2018); Jasanoff (2015, p. 20); Jessop (2010); Moezzi et al. (2017); Otto et al. (2021); Paterson (2021b); Pereira et al. (2019).

Onyebuchi; and short stories by Terh Agbedeh, Mame Bougouma Diene, Osahon Ize-Iyamu, Vuyokazi Ngemntu, Nnedi Okorafor, Suyi Davies Okungbowa, Chinelo Onwualu, Tlotlo Tsamaase, and Jessica Wilson. I also discuss the films *Black Panther* and *Pumzi.*[78] I have chosen these stories because they are some of the most challenging and rewarding for thinking differently about climate politics, and I describe their themes and plots further in Chapter 3.

There are, of course, many other authors of African stories I could have discussed, including Namwali Serpell, Mwangala Bonna, Ebele Mọgọ, Chiagozie Fred Nwonwu, Nisi Shawl, Dilman Dila, Ivor W. Hartmann, Lindsay Redifer, Mike Resnick, Jenny Robson, Sofia Samatar, Ekari Mbvundula Chirombo, Muthi Nhlema, Tade Thompson, Nick Wood, and many others. There are also interesting films I could have included more prominently, like *Crumbs, District 9*, and *Neptune Frost.*[79]

Why include these particular stories rather than others? As already noted, the primary criteria for inclusion is not representativeness, but rather the strength of the challenges offered to dominant ecopolitical imaginaries. Just as theoretical texts are chosen for their utility as tools to answer a question or deconstruct dominant ontologies, these fictional texts have been chosen because of their insights into alternative ways of imagining African climate futures.[80] They are also stories which I (mostly) enjoyed reading and that resonated with my own interests, arguments, and imaginaries or challenged me in productive ways. The selection of these stories, like the wider theoretical approach taken, is fundamentally shaped by my own positionality and worldview and I have tried to be explicit and reflective about this throughout the book. I read these stories at home and I thought about them while walking in the woods, playing with my children and swimming in Stretford and Withington. The photographs of some of these novels in Figure 2 is a reminder that these stories have material form, exist as artefacts and commodities, and were read in particular places. These stories have shaped me and my discussion of them has been shaped by reading them in South Manchester. In reflecting on how this has affected my research I have been influenced by the explosion of work within International Politics on reflexivity, auto-ethnography, and positionality.[81] I discuss this more below.

Some of these stories fit very easily under the label of Africanfuturist climate fiction. Others seem, at least initially, more awkward inclusions. Many of the stories don't mention climate change explicitly and don't dwell on the politics of climate change understood narrowly in the terms set by the UNFCCC

[78] Agbedeh (2013); Beukes (2010); Coogler (2018, 2022); Diene (2022); Ize-Iyamu (2019); Kahiu (2009); Lessing (1999, 2006); Mackay (2022); Ngemntu (2022); Okorafor (2011, 2014, 2015a, 2018,

Figure 2 Photos taken by author

and IPCC. This is a wider question for definitions of climate fiction that is returned to later, but at this point the selection can be justified by reiterating that all these stories have prominent climate and/or environmental themes and are likely to be read by audiences in the twenty-first century in the context of, and awareness of, climate change. The Marvel film, *Black Panther*, for example, is clearly not a film about climate change; yet it does portray an ecologically advanced civilization in Wakanda which has achieved affluence, advanced technology, and apparent harmony with nature, albeit through the extraction of a non-renewable resource. As such the ecological and climate implications of this story have been discussed by critics and reviewers.[82] In another example, Doris Lessing's *Mara and Dann* novels depict a far-future Africa going through a period of drought, while a new ice-age covers Europe. The causes of shifts in the climate are never explained in these novels, neither do they fit within most scientific predictions of future climates. Hence they are not stories about anthropogenic climate change as we understand it. But the stories are explicitly about the ways in which societies adapt—or

2021); Okungbowa (2019); Onwualu (2019, 2022); Onyebuchi (2019); Tsamaase (2016, 2019, 2022); Wilson (2019).

[79] Blomkamp (2009); Llansó (2015); Williams and Uzeyman (2021).

[80] Bleiker (2009, pp. 175–6).

[81] Bleiker (2009, p. 179); Dauphinee (2010); Gani and Khan (2024); Gore (2018); Inayatullah and Dauphinee (2016); Pin-Fat (2016); Ravecca and Dauphinee (2018).

[82] Coogler (2018). See discussion in Abram (2018); Iheka (2021, pp. 1–2); Karam and Kirby (2019); Pereira et al. (2021); Pierre-Louis (2020).

not—to changing climates, and there is a direct critique of the environmentally destructive behaviour of past European civilizations. As such I regard them as part of a broadly defined climate fiction genre, as many other reviewers and critics have done.[83] Other stories discussed in this book foreground oil extraction, floods or drought, and polluted wastelands, in ways that may not directly encompass the greenhouse effect, but which are centrally concerned with the ecological conditions of life in the so-called Anthropocene.[84] There is of course a much wider genre of African eco-fiction—including authors such as Helon Habila, Bessie Head, Ken Saro-Wiwa, Nuruddin Farah, Nadine Gordimer, J. M. Coetzee, and Zakes Mda—which sometimes has a very localized focus on a richly described specific place, whereas I have been drawn to climate fiction in which particular places are located within global contexts. I am also drawn to speculative fiction or science fiction, genres that are again tricky to define but which are future-orientated and operate within, in Darko Suvin's formulation, 'an imaginative framework alternative to the author's empirical environment'.[85]

The question of how 'African' these stories are is another difficult one, inheriting a long history of debates about 'the African novel' and attempts to define and demarcate 'African culture'.[86] There are two points to make here. First, I don't claim that these imaginaries—the policy texts or the fictional stories—represent 'African perspectives' in their entirety, nor of course could they. 'Africa' can be understood in different ways: as a huge continent with great cultural, imaginative, and political diversity; as a signifier for a political identity or movement, such as Pan-Africanism or Africanfuturism; or as a discursive construction.[87] In this book I am most interested in how an imagined 'Africa' is (re)produced and contested—spatially, temporally, ecologically, and politically—by Africanfuturist climate fiction, in ways that challenge dominant discourses about 'Africa' and its place in global climate politics. For these reasons, these stories have been chosen not because they best represent 'African perspectives' but because they challenge dominant ecopolitical imaginaries in illuminating ways.

[83] Lessing (1999, 2006). See discussion in Becket (2009); Johns-Putra (2016); Sargeant (2016); Trexler (2015); Watkins (2010).

[84] For oil extraction, see Okorafor (2011, 2014). For floods or drought, see Agbedeh (2013); Diene (2022); Ize-Iyamu (2019); Ngemntu (2022); Tsamaase (2016, 2022); Wilson (2019). For polluted wastelands, see Beukes (2010); Kahiu (2009); Okungbowa (2019); Onyebuchi (2019); Tsamaase (2019).

[85] Suvin (2016, p. 20).

[86] Ashcroft (2017); Iheka (2017); Julien (2006); Krishnan (2014); Mbembe (2002); Musila (2022); Nfah-Abbenyi (1997).

[87] Mbembe (2001); Okorafor (2019); Tamale (2020).

The second important point to stress here is that it is unhelpful and unnecessary to attempt to assess how 'authentically African' each of these authors and their stories are. Many of these authors—including Okorafor, Kahiu, Beukes, Tsamaase, Okungbowa, and Onyebuchi—have identified themselves or been associated with cultural movements labelled Afrofuturist or Africanfuturist. Other authors included here—such as Doris Lessing—are not usually associated with these movements, although they share many concerns including the role of the past and memory in the present, technology, story-telling, and race and identity. In terms of positionalities and biographies these authors encompass a diversity of complex identities. Nnedi Okorafor writes stories immersed in West African history and culture, informed by her family, upbringing, and numerous visits, but has never lived in Africa.[88] White South Africans, like Beukes, Mackay, and Wilson, are sometimes regarded as more European than African.[89] *Black Panther* is an African-American vision of Africa, in the form of a big-budget Hollywood blockbuster. Perhaps most strikingly, Doris Lessing was British and a member of the literary establishment—a Nobel Laureate, no less—and might justifiably be regarded with some scepticism as an African author. On the other hand, she grew up in Rhodesia (now Zimbabwe) and many of her stories are set in and deeply concerned with African landscapes, characters, and storylines. She is often described as an African writer and she was happy to accept this label whilst refusing to be defined by it.[90] Moreover, the inclusion of white authors like Lessing, as well as Beukes, Mackay, and Wilson, encourage a more direct confrontation with and reflection on issues of Whiteness and Eurocentric cultural power that are central to any discussion of speculative fiction and imagined futures in and from Africa.

It is important to discuss Whiteness and positionality more directly at this point. I am an English-speaking, cis-gendered, white, straight, male academic, writing a book on African political imaginaries and Africanfuturist climate fiction. This book raises questions of whose visions of the future count, who is authorized to narrate legitimate visions of the future, and how those imaginaries challenge or reproduce dominant power relations. I can't avoid turning these questions on myself. Who am I, to write this book? What sort of 'expert' am I?[91] Whose voices do I amplify and whose do I silence, in writing this book? Why write this book at all? These uncomfortable questions have been with me throughout researching and writing. They are raised in the stories and research

[88] Okorafor (2009, p. 276).
[89] Bennett (2018); Tamale (2020, p. 342).
[90] Bertelsen (1986, p. 134); Lessing (2014, p. 18).
[91] Dauphinee (2010, p. 802).

I read; by readers, reviewers, and participants in the conferences and workshops where I have discussed this research; and they have shaped how I have reflected on my own stories and story-telling over the last five years. There are no easy answers to any of these questions. But writing a book—whether fiction or not—is a form of 'self-writing', in which a narrative is crafted that brings into being both a self and a world.[92] By this I mean that writing can offer glimpses of fractured 'short stories of selves', continually dissolving and reforming, rather than culminating in a coherent and unified subject.[93]

Who am I, to write this book? I read a great deal of science fiction as a teenager—escaping into the worlds of Isaac Asimov, Frank Herbert, Ursula Le Guin, H. G. Wells, and Terry Pratchett on my father's bookshelves—and returned to science/speculative fiction much more recently to discover a very different set of authors, visions, and politics. For circumstantial reasons, my discovery of Africanfuturist climate fiction became intertwined with my research. After our second child was born I had just enough energy to read fiction as I held a sleeping baby. When it was nice weather I walked in the woods with my daughter. When it was wet I could read a slim paperback or online story as I rocked her to sleep inside. During a period in my life when foreign travel or fieldwork seemed unimaginable, reading fiction was one way to explore distant worlds. As I returned to academic work after shared parental leave I wanted to reflect further on, and share, what I had read. In the chapters that follow I continue to discuss identity, narratives, positionality, and reflexivity, but at this stage a few more comments on race and Whiteness are required.

In this book I use 'Whiteness' to mean more than pale skin colour. Rather, Whiteness is a set of ontological and epistemological structures, a 'system of privilege' for those 'of a phenotypical white presentation' that is ultimately built upon the 'mythologised historical genius' of White/Eurocentric civilization.[94] Whiteness exerts its power through its invisibility, in that discussions of race, racism, and racialization are usually assumed to be about Others; but conversely, the grand narratives of modernity, civilization, capitalism, and secularism are implicitly underpinned by Whiteness. Whiteness is, for Clive Gabay, 'a constitutive element in the construction of imaginations about international order'.[95] In particular, and crucially for this research, visions of alternative climate futures often raise explicit or implicit questions about the

[92] Mbembe (2002, p. 272).
[93] Pin-Fat (2016, p. 33).
[94] Gabay (2018, pp. 2, 9).
[95] Gabay (2018, p. 4). See also Baldwin (2012); Mitchell and Chaudhury (2020); Musila (2014).

perpetuation or collapse of Whiteness, including the sustainability of Western high-consumption lifestyles, climate refugees and migration, population growth and/or decline, reparations for past harms, and the prospects for global governance under a liberal international order.[96]

As someone of a phenotypical white presentation, working within a historically white institution and a discipline structured by Whiteness, I benefit from these structures even when I try to disavow them in my work. This necessitates careful consideration of the ethics and risks of research. For example, there are real risks of appropriation and reinforcing unequal power relations when 'someone like me' writes a book on this topic. Does a white, European, male academic writing about African climate futures risk further marginalizing and decentring African voices? I have thought a lot about this and discussed it with colleagues, students, readers, and reviewers. There are no easy answers and ultimately, yes, the risk is real. As Julia Gibson puts it, there is an inescapable 'tension of working with' stories from Indigenous or Afrofuturist traditions, 'especially for those of us coming from environmentally privileged and/or settler backgrounds'.[97] Nevertheless, I decided to write this book. Several reasons have motivated this choice. First, given the degree to which African climate imaginaries remain marginalized in discussions of global climate politics and fictional futures, I hope that this book can play a role in opening up more space for, and amplifying, African perspectives, voices, and stories. I have explicitly chosen stories with anti-colonial, anti-racist, anti-capitalist, and anti-patriarchal themes, and writing this book is a political intervention that is intended to challenge some of the wider structures of power within which I am situated.

Second, I have tried to practise a mode of reading which is care-full, thoughtful and reflexive. I started from a position of curiosity and humility; there is a great deal I don't know about the wider context and cultural background of these stories. I also make no judgements about the aesthetic or literary merits of the fiction, but rather read these stories as reflections of and challenges to global ecopolitical imaginaries. I have read each story several times and I have read non-fiction essays and listened to interviews with many of these authors. I have immersed myself in some of the extensive secondary literature on them. I aim to use these stories to support arguments that seemed to me to be broadly in alignment with the ecopolitical outlook of the authors, and more importantly I believe that engaging with them seriously and

[96] Oreskes and Conway (2014); Paterson (2021a); Sabaratnam and Laffey (2023); Sultana (2022); Sylla et al. (2024); Táíwò (2022); Whyte (2018).

[97] Gibson (2020, pp. 208, 219). See also Bennett (2018).

in detail is no less than what they deserve as powerful contributions to wider discussions about global climate politics.[98]

Third, I acknowledge that there is an irreducible element of appropriation and extraction in this project. I am using these stories to inform my own arguments about global climate politics and ecopolitical imaginaries. Fictional stories are not usually written for these purposes and the authors may well disagree with many of my arguments or reject this project entirely. I also acknowledge my limitations in terms of only engaging with English-language imaginaries in this book, posing further challenges in terms of trying to explore counter-hegemonic imaginaries through the hegemonic language of Anglo-American empire.[99] Yet the power relations between me and these authors—and within and among these authors—are not always straightforwardly hierarchical, even if they exist within global structures of power that remain racialized and gendered.[100] Most of these authors are successful, professional artists, including winners of numerous awards and prestigious fellowships. Moreover, being a published author entails opening one's work to critique and alternative interpretations. The political strategies and fictional tales examined in this book are all public texts, intended to be read and discussed as widely as possible. I hope that, as Rahul Rao puts it, engaging with fictional texts entails 'less invasive forms of entry into worlds that are not one's own.'[101] That said, all acts of writing and critique are political and potentially violent. This book is an explicitly political intervention and is also open to response and critique.

Moving on, provisionally, from issues of positionality and reflexivity to questions of method and structure, in Chapters 4–7 I discuss Africanfuturist climate fiction in terms of how these stories challenge the spatiality, temporality, ecology, and politics of dominant ecopolitical imaginaries. This entails asking questions about where they are set, when the story takes place, whom the story is about (including other non-human species), and what political implications I draw from them. These questions emerged iteratively from reflection upon dominant narratives and the gaps in both official political strategies and the wider genre of climate fiction in North American and European stories.[102] The stories encompass the transformation of Lagos into a post-fossil fuel utopia; far-future epics on continental scales

[98] Alberro (2024, p. 186).
[99] Bleiker (2009, p. 171); Said (1994, p. 270).
[100] Gani and Khan (2024, p. 6).
[101] Rao (2014, p. 202).
[102] Milkoreit (2016, p. 178).

where post-apocalyptic civilization survives in Africa; fantastical tales of new relationships between humans and nonhuman entities set in Johannesburg, the Western Cape, KwaZulu-Natal, Botswana, and the Niger Delta; revenge fantasies; utopian dreams of bucolic homesteads and reparative communities; and dystopian warnings about the dangers of digitization and rampant urbanization.

The ground is prepared for this analysis in Chapter 3, 'Why read African climate fiction?' Here I unpack the case for why we should read climate fiction as contributing to climate theory. What do stories *do*, politically? In responding to this question, I draw upon a growing body of political theory that reads and invokes fictional texts and I illustrate the argument with Nnedi Okorafor's novel, *Lagoon*.[103] I provide a brief introduction to the burgeoning field of climate fiction more generally, before exploring the role of stories from and about Africa. In contrast to the historic and recent prominence of African stories and authors in world literature more generally, African settings and characters remain relatively marginal in science fiction and climate fiction, genres themselves which are still often dismissed as 'low-brow' or popular by advocates of realist literature.[104] Yet, there is a growing body of Africanfuturist climate fiction and these stories have an important part to play in challenging global ecopolitical imaginaries.[105] The final section of the chapter introduces the authors and stories that the rest of the book discusses in more depth.

Chapter 4, 'Strange Places and Unfamiliar Homesteads', explores where African climate fictions are located. Dominant climate imaginaries have tended to focus on European and North American cities and coastlines or melting polar glaciers. African landscapes are rarely prominent in global imagined climate futures, except where contemporary famines and desertification are held up as a forewarning of global futures.[106] Even in African long-term strategies the preeminent spatial representation of the future is urban and technologized, often focused on energy infrastructure like windfarms and solar plants. These dominant imaginaries are challenged in this chapter through the techniques of estrangement and defamiliarization found in a range of Africanfuturist climate fiction novels, films, and short stories. Three particular features of place are identified. First, these stories are often located in unsettling,

[103] Examples of political theorists using fiction include: Bell (2021); Bleiker (2009); Edkins (2013); Hunt (2023); Jackson (2020); Ling (2014); Mitchell and Chaudhury (2020); Odysseos (2019); Opondo (2022); Rao (2014); Shapiro (2010); Thaler (2022); Weheliye (2014); Welland (2018); Whyte (2018); Zamalin (2019). See also Death (2022b).

[104] Trexler (2015, p. 6).

[105] Adejunmobi (2016).

[106] Johns-Putra (2019); Oreskes and Conway (2014); Trexler (2015).

unfamiliar landscapes that can be described as heterotopian in the way that they disrupt and disturb our maps of the world.[107] Second, the focus on cities in many of these stories works to destabilize utopian and hyper-modernist 'African urban renaissance' storylines.[108] Third, many of these stories pay close attention to the domestic spaces and homesteads of the future, often suggesting that families and homes are places of danger and insecurity.[109] There are some powerful political implications to be drawn from these three themes, implications that can contribute to important discussions about the spatiality of climate change politics and low carbon transitions, and that can help to draw maps that might orientate us better in a climate-changed world.[110] African-futurist climate fiction can lead us to demand climate strategies that include people and places more prominently, and envision forms of city-hinterland that are fluid, porous, and in-flux, as well as radically transformed domestic environments and care-relations that are imaginative and hospitable.

This chapter takes us to a wide range of African landscapes and settings: from Johannesburg's sewers to skyscrapers in utopian Lagos, high-speed railways, domed-enclaves, and solar-powered barges on the Congo, futuristic East African compounds to homely eco-villages in the Niger Delta, and deserts from the Sahara to the Kalahari. Particularly close attention is paid to Okorafor's novel, *Who Fears Death*, Tsamaase's short story, 'Virtual Snapshots', and Mackay's novel *It Doesn't Have to Be This Way*. These stories greatly enrich our abilities to envision the places of climate-changed futures.

Chapter 5, 'Heterotopian Temporalities and Afrocentric Futures', asks when are African climate futures set and how do they challenge mainstream understandings of time and temporality? Dominant socio-climatic imaginaries tend to offer linear visions of homogenous time. The usual way in which time is represented in climate change narratives—both scientific/policy and fictional—is in terms of the pressing urgency of action, a sense that time is running out for humanity. This chapter argues that a core feature of Africanfuturist climate fiction is that the passing of time is rendered more complex and more unpredictable than in dominant climate temporalities. Africanfuturist climate fiction challenges dominant temporalities in four ways: by imagining Afrocentric global futures; through heterotopian temporal structures that interweave past, present, and future in multiple ways; by insisting on the active role of the detritus of the past in the present and future, alongside an insistence on the

[107] Jameson (2007, p. 144).
[108] Hoelscher et al. (2023); Iheka (2021).
[109] Johns-Putra (2019).
[110] Thaler (2022).

possibility of utopian futures; and the predication of all this on the inevitability of death, decay, and suffering.[111]

In Chapter 5 I discuss a range of novels, short stories, and films from across the genre of Africanfuturist climate fiction. The novels of Nnedi Okorafor, especially *Lagoon* and *Who Fears Death*, are particularly important for challenging dominant temporal narratives. The *Black Panther* films are also significant in imagining an Afrocentric global future and Wanuri Kahiu's short film *Pumzi* narrates heterotemporal futures in interesting ways. This chapter also engages closely with the *Mara and Dann* novels by Doris Lessing, set in a far-future 'Ifrik', and reflects on the ways in which fiction can both reproduce and challenge wider structures of Whiteness in climate change politics and culture. Novels, films, and short stories can—like museums and art galleries—work to defamiliarize and estrange us from conventional narratives about our past, present, and future, and this is an important way to begin to repair the damage wrought by colonialism, racism, patriarchy, and other injustices. In this chapter I argue that there should be more space for remembrance, grief, mortality, and compost in dominant climate imaginaries.

In Chapter 6, 'Dangerous Ecologies and More-than-Human Conviviality', I ask: who is the story about? This chapter pays particularly close attention to human relationships with non-humans. Dominant climate imaginaries rarely devote much attention to future ecological relationships, except by noting the increasing importance of conservation for threatened species and ecosystems. In contrast, science-fictional and speculative imaginaries have proved fertile terrain for telling more-than-human and multispecies stories. This is especially true of authors like Lauren Beukes and Nnedi Okorafor, whose wild lifeworlds conjure new human–nonhuman relationships that challenge the resolutely human-centred imaginaries of policy strategies. In this they echo Achille Mbembe's observation that '[b]eing human no longer determines the limits of those occupying the world'.[112]

This chapter argues that climate politics and low carbon development visions are impoverished by the absence of more-than-human politics. It shows how Africanfuturist climate fiction can make an important contribution to reimagining co-dependencies between humans and animals, ethical bonds between people and forests, deserts and waters, and encounters with nonhuman spirits or posthuman alien civilizations. This argument is made in three parts. First, the question of how to relate well and engage ethically

[111] On alternative temporalities, see Granjou and Salazar (2016); Grosz (2004); Hugo (2017); Jackson (2020); Mitchell and Chaudhury (2020); Weheliye (2014); Whyte (2018).

[112] Mbembe (2019, p. 13). See also Ghosh (2016); Haraway (2016); Melzer (2006); Mitchell (2023).

with nonhuman others is prominent in most if not all the stories examined in this book. Second, many of the Africanfuturist stories discussed here entail the emergence of quite radically new forms of human subjectivity. Third, these changing relationships and changing subjectivities produce new visions of living with radically different others. Africanfuturist fiction is profoundly convivial, concerned with ways in which a more-than-human 'we' might rub along together in the future, but also insistently cognisant of the trouble that this will bring. Climate change has brought challenges of finding ways to die well together, as well as lively flourishing, and this will involve sacrifices and attempts to make reparations for ecological harms already done.[113] The chapter discusses Okorafor's novel *Lagoon* and Beukes' novel *Zoo City* in detail, as well as a number of short stories and films, in order to argue that embracing human difference—racial, gendered, sexual, differently abled, neuro-diverse—will be vital in coming to terms with changing climates. Moreover, the search for ways of living and dying well alongside nonhuman others will involve imaginative and radical modes of 'interspecies diplomacy' as well as reconsideration of what humanity itself entails.[114]

Chapter 7, 'Reparative Climate Politics and Reimagined Subjectivities', returns to the over-arching question of the political implications of these stories. While visions of the future are inescapably political, dominant policy imaginaries are often technical, institutionalized and depoliticized. They rarely ask: who are the winners and losers? Will the process of change be democratic and how will it overcome entrenched vested interests? Science-fiction stories, on the other hand, offer freer rein to imagine the subjects and subjectivities of the future, the winners and losers, and more violent or revolutionary processes. Reading fiction can make the personal political by helping us to imagine what it is like to stand in another's shoes.[115]

The chapter begins by observing the prominence of revolutionary and revenge narratives in these stories, from blockbuster films like *Black Panther* to Okorafor's novels *The Book of Phoenix* and *Noor*. In this respect these stories have potential lessons for climate activists in learning 'how to fight all over again'.[116] These stories embody the fragile, utopian insistence expressed by Lessing's character Anna Wulf in *The Golden Notebook*: 'If people can imagine something, there'll come a time when they can achieve it.'[117] In this spirit I go on to ask: what comes next? What comes after the revolutionary upheaval

[113] Haraway (2016); Iheka (2017).
[114] Constantinou and Christodoulou (2024); Fougner (2023); Opondo (2022); Strausz (2024).
[115] This role of fiction has been most famously expressed by Nussbaum (2017).
[116] Malm (2021, p. 22).
[117] Lessing (2014, p. 248).

portrayed in these stories? There are powerful reparatory themes in many of these novels, lending imaginative heft to demands to pay climate debts, make ecological reparations, and heal loss and damage.[118] These stories can help us think through what a radical, reparative politics might entail, and what sort of future societies might lie ahead. Africanfuturist stories include hypermodernist ecopolitical imaginaries of science and technology-driven rampant capitalism, neo-feudal tales of warlords and chieftains, and glimpses of post-capitalist political economies.[119] In contrast to many other speculative climate fiction novels, however, the greatest strength of these stories lies in their envisioning of new eco-hero(ine)s with the power to reimagine the past and rewrite the future.

The third section of Chapter 7 develops this argument. Many of these stories feature marginal, queer, subaltern, feminist, outcast heroines, posthuman actors, and radical black utopians who respond to Fanon's call to 'introduce invention into existence'.[120] This section highlights the prominence of radical feminist, queer, and cyborg characters in stories like *It Doesn't Have to Be This Way*, *Noor*, 'Spider the Artist', and the *Mara and Dann* fables. These transgressive characters are some of the most powerful contributions of Africanfuturist climate fiction, but they must also be read with one eye on the risk of over-individualizing how we imagine political change.[121] Significantly, a repeated form of political agency in these stories is the power to write and rewrite stories themselves. This brings us back to a central question for the book: why read (or write) climate fiction at all?

In the book's final chapter, 'Rewriting the ending', I set out my overall answer to this question and draw together the main political implications that arise from these stories. In coming to an ending, Chapter 8 also reflects upon the ways in which some of the authors studied here have ended their stories in particularly memorable and striking ways. Okorafor's *Who Fears Death* and *The Book of Phoenix* both conclude with reflections on the process of telling, writing, and rewriting stories. Other stories by Kahiu, Lessing, Mackay, and Onwualu also end by foregrounding impermanence, memory, memorialization, and the death of the author. Literary luminaries like Ben Okri and Doris Lessing have been haunted by the fear that maybe our stories are all that will survive of humanity in a post climate-cataclysmic future.[122]

[118] Patel and Moore (2017); Sze (2020); Táíwò (2022).
[119] Robinson (1992, 1993, 1996, 2017, 2020). See also Hatzisavvidou (2024); Schneider-Mayerson (2019).
[120] Fanon (2008, p. 180).
[121] MacGregor (2014); Maniates (2001).
[122] Okri (2023, p. 42); Lessing (2007).

Ultimately, inspired by Okorafor's character Onyesonwu in *Who Fears Death*, I suggest that we need to try to live in some of these worlds that we've imagined. After all, as the song goes, the sun doesn't go down, it's only an illusion caused by the world spinning around. Reading stories is not an alternative to political action or an escape from politics. Connecting these stories to more direct forms of political action—the drafting of policy strategies, building community organizations, holding the powerful to account—is one way to mitigate against the 'delusions of self-importance' that Clark sees in some ecocriticism.[123] As texts alone, these policy strategies and fictional stories are not necessarily politically significant; rather it is only when they are read in a certain way, or discussed and critiqued and otherwise enacted, that they become political. We can use these stories to fortify and inform our political interventions, enacting the best aspects of these stories or rewriting them to craft new stories. As Grosz suggests, our most urgent political task is not to predict the future but to invent it.[124]

Overall, this book argues that imagined futures are absolutely central to the global politics of climate change, and that African ecopolitical imaginaries have been hitherto marginalized but are vital to consider in the struggle for just, equitable, and sustainable climate futures. The stories discussed in this book can inspire more radical, reparative, more-than-human ecopolitical imaginaries, by helping to weigh the debts we all owe, imagine what reparations might entail, and explore the contours of living convivially alongside more-than-human others in heterotopian, climate-changed futures. While official political strategies and visions of the future deserve to be taken more seriously in political economies of transition, the failure of existing climate strategies and debates about energy transitions and green new deals to confront questions of land and climate justice, loss and damage, extractive political economies, and the racialized and gendered violence of capitalism and colonialism is a fatal flaw. The current world order confines too many people—human and nonhuman—to lives of violence, inequality, and injustice, and if we want to survive to tell new and better stories in more peaceful, egalitarian, and just futures then we need more radical ecopolitical imaginaries.

[123] Clark (2015, p. 198). Dismissive perspectives on the political potential of writing and reading fiction can be found in Pattee (2023) and VanderMeer (2023). For a more imaginative insistence on the links between fiction and political action, see Imarisha and Brown (2015).

[124] Grosz (2004, p. 261). See also Edkins (2013, p. 283) and Vervoort et al. (2015, p. 66).

2

Dominant Narratives in Long-Term Strategies

Net zero targets have been announced by 147 countries worldwide, covering 88% of global GHG emissions.[1] This might appear to suggest that there is broad agreement to strive for a climate-secure future through the mitigation of greenhouse gas emissions. Yet, at the same time, no oil, gas, and coal producer countries have committed to reduce fossil fuel production in line with what is necessary to limit warming to 1.5°C, and the world is on track to produce more than double the amount of fossil fuels in 2030 than would be consistent with this target. COP28 in December 2023 failed to agree on a phase-out of fossil fuel use as many had demanded and hoped, instead agreeing to 'transition away from fossil fuels'.[2] Net zero targets in themselves mean very little without robust budgets for greenhouse gas emissions and commitments to plausible political strategies for achieving mitigation, adaptation, reparation, and climate justice. There are few signs that the world is on a pathway to a sustainable and secure climate future. Indeed, it seems that 1.5°C of global warming has already been breached.

Such a bleak situation can lead to various emotions, including fear, sadness, anger, despair, or a bleak sense of humour. My own emotional responses to climate catastrophe encompass all these. I fear for the future, especially for the world my children will grow up in, and the current scale of violence and injustice in the world makes me angry. I also feel a sense of guilt and responsibility, owing to my relatively privileged, high-consumption lifestyle, made possible not least by my father's life-long career for a large petrochemical corporation. I have responded by marching for climate justice, working for Greenpeace, teaching critical environmental politics, standing on picket lines, and conducting research on environmental justice movements in South Africa and elsewhere. In this research I want to subject dominant stories about climate futures to critical scrutiny and search for more radical stories about climate

[1] Net Zero Tracker (2024). See also Green and Reyes (2023).
[2] Shirley et al. (2023). See also SEI et al. (2023).

African Climate Futures. Carl Death, Oxford University Press. © Carl Death (2025).
DOI: 10.1093/9780198960775.003.0002

politics. In subsequent chapters I look at fictional stories about African climate futures, some of which are terrifying, others are hopeful, and many are bitingly satirical. This chapter looks at the narratives of climate futures in official African government strategies and policies. There appears to be little to joke about here. But the incongruous gap between proclaimed ambition and tangible implementation means that sometimes we can only laugh. A queer, ironic, dissident, irreverent sensibility is advocated by Seymour in what she terms 'bad environmentalism'.[3]

In this spirit, I want to start this chapter on African long-term low carbon strategies with a joke Ebenezer Obadare retells about Nigerian politics:

> A little boy goes to his father and asks, 'What is Nigerian politics?' Dad says, 'Well, son, let me try to explain it this way:
>
> A. I am the head of the family, so call me the President.
> B. Your mother is the administrator of the family money, so we call her the Government.
> C. We are here to take care of your needs, so we'll call you the Nigerian People.
> D. The nanny, let us regard as the Working Class.
> E. And your baby brother, we'll call him the Future. Now think about that and see if it makes any sense.'
>
> So the little boy goes off to bed thinking about what Dad has said. Later that night, he hears his baby brother crying, so he gets up to check on him. He finds that the baby has severely soiled his diaper, so he goes to his parents' room and finds his mother sound asleep. Not wanting to wake her up, he goes to the nanny's room, and finding the door locked, he peeks through the keyhole and sees his father in bed with the nanny. He gives up and goes back to bed. The next morning, the little boy says to his father, 'Dad, I think I understand the concept of Nigerian politics now.' The father says, 'Good, son, tell me in your own words what you think Nigerian politics is all about.' The boy replies, 'The President is screwing the Working Class, while the Government is sound asleep. The People are being ignored and the Future is in deep shit.'[4]

The gendered, sexual, and classed connotations of this sort of joke invite critical scrutiny, but in terms of climate politics it is hard to avoid the conclusion

[3] Seymour (2018). See also Boykoff and Osnes (2019); Mbembe (2001, p. 102); Musila (2014).
[4] Obadare (2009, pp. 256–7).

that the future is in deep shit. There is a massive gap between stated commitments and aspirations for the future—keeping global temperature rises to under 1.5°C or 2°C, decarbonizing economies, ensuring sustainable consumption and production patterns, achieving climate justice, including through addressing loss and damage and making reparations—and the actual policies and implemented measures taken by governments and other bodies worldwide.

This is the broader context for the task of this chapter, which examines how African climate futures are narrated by looking at some of the dominant political visions of long-term climate strategies. Given the deeply concerning forecasts about the impacts of climate change on Africa—often described as one of the world's most vulnerable regions to climate change—there is widespread agreement that there needs to be a global transition to a decarbonized global economy, with particular emphasis on low carbon and climate resilient development models in the Global South. Academic literature on sustainability transitions is at the forefront of research in this field and there is a growing recognition that state-led planning and strategies will play a central role. Accordingly, studying the ways in which climate futures are imagined, envisioned, and narrated by state actors is crucial.

For this reason, I focus in this chapter on the official Long-Term Strategies (LTS) submitted by governments to the UNFCCC under the 2015 Paris Agreement, which are intended to set out a pathway for decarbonization and institutionalize forms of mitigation necessary to achieve the Paris goals. These, I argue, can be regarded as 'present futures', policy enactments of ecopolitical and socio-climatic imaginaries, and are therefore dominant in the sense that they are institutionally authorized and mandated by sovereign states, legitimized by their status as official submissions to the preeminent international institution for governing climate change.[5] This is not to claim that they are dominant in the sense of being the most powerful imaginary of political-economic futures in the countries concerned. In contrast, they sometimes feel like little more than a sick joke. Net zero visions of climate justice, low carbon development, and ecological resilience contrast strikingly with the Nigerian government's determination to maximize oil extraction for the foreseeable future, the South African economy's dogged attachment to coal, ongoing civil war in Ethiopia, a crisis of democracy in Zimbabwe, and donor dependence in The Gambia. Yet, these countries, together with Benin and Morocco, were

[5] Adam and Groves (2007, p. 28). See also Hatzisavvidou (2024) and Milkoreit (2017). This chapter draws on and expands the discussion in Death (2024a).

the first African countries to submit LTS to the UNFCCC, and they have claimed a role of climate leadership and ambition in Africa.[6] Their visions of climate futures deserve to be taken seriously and subjected to critical scrutiny, even if the stated plans might seem closer to science fiction than political fact. Reading, scrutinizing, and analysing these strategies stems from a wider belief that greater democratic engagement and contestation over imagined climate futures is necessary for more effective and just forms of climate politics.[7]

The chapter discusses literature on transitions to low carbon futures and the role of ecopolitical imaginaries in shaping these transitions. Drawing on research on narratives in policy texts, it sets out four key sets of questions for these strategies: what are their plots and policies; what are the lead actors and supporting characters; what are the main discourses and themes; and what are some of the gaps and silences in these long-term strategies? In this chapter I argue that there are important gaps in terms of the overarching visions of what a decarbonized economy would look like; how to get there, in terms of overcoming obstacles and enemies; the role of wider social engagement and democratic participation; the types of political subjects and agents imagined as driving the transition; and an ecological perspective that goes beyond anthropocentrism. None of these strategies makes a strong case for climate reparations or the climate debt owed by the Global North to the Global South. The rest of the book seeks to address these gaps. I argue that Africanfuturist climate fiction can play an important role in the political task of strengthening radical visions of climate futures.

Transitions to Low Carbon Development

Climate change is not simply a future prospect, especially in Africa. The chapter on Africa in the sixth assessment report of the IPCC in 2022 states with high confidence that 'key development sectors have already experienced widespread losses and damages attributable to human-induced climate change, including biodiversity loss, water shortages, reduced food production, loss of lives and reduced economic growth'.[8] The IPCC predictions for the future are even gloomier, with negative impacts of climate change in Africa

[6] On Benin's strategy, see Ross and Fransen (2017).

[7] For different ways of articulating this argument, see Bai et al. (2016); Moore and Milkoreit (2020); Newell (2021); Pereira et al. (2019).

[8] Trisos et al. (2022, p. 1289).

becoming more 'widespread and severe'.[9] Droughts will become more frequent and, if global temperatures rise by 3°C or more, they could last twice as long in parts of North Africa, the western Sahel, and southern Africa. Flooding will increase and hundreds of millions will be exposed to sea level rises.[10] The glaciers on Mount Kilimanjaro will be gone by 2040. Crop yields will be reduced for staple foods across most of Africa. Damage to infrastructure and loss of biodiversity will increase rapidly. If global temperatures rise by between 2.5°C and 4°C, the annual number of days above potentially lethal heat thresholds in tropical Africa is likely to reach 200–300 during the twenty-first century. Tens of millions more people will be exposed to vector-borne diseases. Temperature rises of 2.5°C could lead to 56–86 million people migrating internally in sub-Saharan Africa, due to water stress, reduced crop productivity, and sea level rise.[11] These dire forecasts contribute to a global context in which 'the possibility of a major collapse of global human civilisation can no longer be discounted, within living timeframes'.[12]

The pictures created by these stark figures is one reason why most African leaders and governments are enthusiastic proponents of ambitious global mitigation, adaptation, and climate finance targets. African strategies and politicians echo the IPCC in pointing out that Africa is one of the most vulnerable regions to climate change but also one of 'the lowest contributors to greenhouse gas emissions'.[13] Despite highly differentiated responsibilities, most African leaders acknowledge that their countries will have to play their part in climate mitigation and transitions to a decarbonized economy, although there are intense debates and differences over the use of oil, gas, and coal reserves during the transition; the relative priorities afforded to economic growth, poverty reduction, and decarbonization; and the role of international finance to compensate for loss and damage and pay for the transition to low carbon development.[14]

It remains the case that much of the academic literature on low carbon transitions has focused on the developed economies of the Global North; however, there is an increasing volume of academic research and political impetus behind sustainability transitions in the Global South, including in Africa. There are many different academic approaches to transitions. These include the Multi-Level Perspective on the opening of renewable energy

[9] Trisos et al. (2022, p. 1289).
[10] Trisos et al. (2022, pp. 1290–1).
[11] Trisos et al. (2022, pp. 1291–2).
[12] Paterson (2021a, p. 397).
[13] Trisos et al. (2022, p. 1289).
[14] Okereke (2024).

'niches' within broader energy regimes and policy landscapes; research on foresight and pathway scenarios; work on the 'seeds' that might lead to 'Good Anthropocenes'; and on 'socio-technical imaginaries' that might open paths to alternative futures.[15] Many critical political economists call for more radical forms of political-economic transformation.[16] One shared feature of these literatures is the recognition that state-led national planning is back in vogue, this time in the name of structural transformation and decarbonization of the global economy.[17]

State-led processes of economic transformation require accompanying visions, ideologies, and imaginaries to legitimate and communicate them. As Manjana Milkoreit puts it, '[e]xplicit visions of desirable (sustainable) and undesirable futures are necessary to motivate and guide any kind of change, but might be particularly important for triggering transformational change'.[18] Political scenarios and visions are a 'narrativisation of future possible worlds', and thus it is worthwhile to interrogate these visions in order to assess the role they play in both reflecting and shaping dominant power relations.[19]

Long Term Strategies as Ecopolitical Imaginaries

Parties to the UNFCCC—as well as cities, businesses, communities, and others—have come up with a range of strategies and visions, each of which set out a particular vision of the future, more or less explicit, and a set of policies or values to get there. The core mechanisms of the 2015 Paris Agreement are Nationally Determined Contributions (NDCs), which detail the efforts by each country to reduce national emissions and adapt to the impacts of climate change. The Paris Agreement also stated that all countries 'should strive to formulate and communicate' mid-century, long-term, low greenhouse gas

[15] On the MLP, see Bridge et al. (2013); Feola (2020); Geels (2014); Newell and Phillips (2016); Okereke et al. (2019); Osunmuyiwa et al. (2018); Osunmuyiwa and Kalfagianni (2017); Power et al. (2016); Swilling et al. (2016). On foresight, scenarios and pathways, see Kanitkar et al. (2024); Okereke (2021); O'Neill et al. (2017); Pathak (2017); Pulver and VanDeveer (2009); Trollip (2021); Vervoort et al. (2015); Vervoort and Gupta (2018); Williams and Waisman (2017). On 'good Anthropocenes', see Bennett et al. (2016); Hamann et al. (2020); Pereira et al. (2018). On socio-technical imaginaries, see Cairns et al. (2022); Jasanoff (2015); Simmet (2018); Tozer and Klenk (2018).

[16] Albert (2024); Moore and Milkoreit (2020); Newell (2021); Okereke (2024); Patel and Moore (2017); Paterson (2021a, 2021b); Pearse (2021).

[17] Alami et al. (2021); Chimhowu et al. (2019); Eatough (2015, 2017); Newell (2021, p. 93); UNDP (2017).

[18] Milkoreit (2017, p. 1). See also Adam and Groves (2007, p. 29); Jessop (2010); Levy and Spicer (2013); Newell (2020, 2021); Paterson (2021b, p. 924); Pereira et al. (2018); Pulver and VanDeveer (2009); Rennkamp (2019); van der Helm (2009, p. 99); Vervoort and Gupta (2015); Wright et al. (2013).

[19] Eatough (2015, p. 701).

emission development strategies by 2020.[20] These were intended to be longer-term strategies than the NDCs, broader in scope, which could narrate 'robust visions of the low-carbon futures of countries'.[21] Advocates of LTS urge that they should set-out 'pathways to visualize their transition to low-GHG economic development' and incorporate 'a process to expose significant political economy issues for the entity undertaking the analysis (who wins, who loses, and what can be done about it)'.[22] Therefore, at least in theory, these LTS go beyond narrow, technical documents, and offer countries an opportunity to articulate a broader vision of a low carbon, climate resilient society of the future, together with pathways to get there.

By the end of 2023, 68 countries had submitted LTS to the UNFCCC, including many of the largest economies in the world. Submissions continued well past the 2020 target date, with nearly a fifth of total submissions coming in 2023. However, there are fewer submissions from developing countries, very few submissions from Africa, and few from fossil-fuel producing and exporting developing economies. Nigeria was the first and only OPEC member to make a submission before the end of 2023.[23]

Although there is a lack of in-depth research on the LTS so far, it is possible to observe many shared features and similarities across the strategies. This reflects shared assumptions about the sorts of pathways necessary to reach the Paris goals, including a global shift to renewable energy, greater energy efficiency in transport, construction, industry and everyday consumption, and expanding carbon sinks like forests.[24] The national strategies make the case for the role of government planning: as the South African strategy puts it, 'failing to plan is planning to fail'.[25] They usually describe the consultative processes behind the strategies and the importance of wider public engagement with the transition process. As Admos Chimhowu et al. conclude, we are now in 'a new era in which plan documents are no longer an end in themselves but are seen as part of a process of communicating and negotiating national ideals with internal and external audiences'.[26]

These long-term strategies must be viewed critically and they invite scepticism. My starting assumption is that they are neither completely meaningless

[20] UNFCCC (2023). See also Tobin et al. (2018).

[21] Pathak (2017, p. 3).

[22] Pathak (2017, pp. 2–3).

[23] UNFCCC (2023). Nigeria submitted a Long-Term Vision in 2021 and a Long-Term Strategy in 2023. See Federal Ministry of Environment (2021) and National Climate Change Council (2023).

[24] Death (2024a); Pathak (2017); Ross and Fransen (2017); Ross et al. (2021); Williams and Waisman (2017).

[25] Republic of South Africa (2020, p. xvii).

[26] Chimhowu et al. (2019, p. 86).

nor are they the masterplans for a structural transformation on the scale needed to tackle the climate crisis. They do have a certain power although their influence shouldn't be overstated. The LTS frame certain issues as priorities and particular pathways as desirable; they render other issues invisible; and they acquire a significant status as a text that can be invoked or contested within increasingly intense political battles over decarbonization. They are often aimed at attracting financial support and investment for green transition policies, especially for developing countries. On the other hand, a convincing strategy document does not mean that it can or will be implemented, nor that it will have the intended effects. References to public participation and consultation do not mean there has been a genuine process of social dialogue. Newell et al. cast doubt on the degree to which green transitions are 'amenable to the rolling out of preformed "master plans" or "blueprints"'.[27] For literary theorist Matthew Eatough, '[t]he planning document thus becomes a sort of fetishistic substitute for social action, the drafting of its speculative "vision" tantamount to the creation of that future'.[28]

In this book I am not directly concerned with the implementation of the plans or even in assessing how 'realistic' they are, although as this chapter noted at the outset, the gap between visions and current actions is the inescapable context for any discussion of climate politics. Rather, in the rest of this chapter I propose to take these strategies on their own terms, as texts that reflect wider ecopolitical imaginaries. As discussed in the previous chapter, imaginaries are 'collectively held, institutionally stabilized, and publicly performed visions of desirable futures', which 'encode what is attainable and envision how life ought (or ought not) to be in the future'.[29]

It is important to examine the ways in which these strategies narrate an energy transition to a low carbon future. An inspiring narrative and vision are often argued to be crucial for convincing audiences about the desirability and feasibility of transition, and Williams and Waisman argue that the 'creation of narratives describing potential routes to a low-carbon future' is central to effective long-term strategies.[30] Drawing on literature on narratives in policy texts, there are four key sets of questions to ask when exploring how LTS narrate energy transitions.[31] First, what is the overall plot, and what policies do they

[27] Newell et al. (2021, p. 905). See also Newell (2021, pp. 144, 203–4); Paterson (2016, 2021b).

[28] Eatough (2017, p. 248).

[29] Jasanoff (2015, p. 4) and Marquardt and Nasiritousi (2022, p. 624), respectively.

[30] Williams and Waisman (2017, p. 14). See also Jasanoff (2015); Levy and Spicer (2013); Newell (2021, p. 203); Pereira et al. (2019); Ross and Fransen (2017); Ross et al. (2021); UNDP (2017).

[31] Beck (2018); Jasanoff (2015, p. 20); Jessop (2010); Miller et al. (2015); Moezzi et al. (2017); Otto et al. (2021); Paterson (2021b); Pereira et al. (2019); Raven (2017); Vervoort et al. (2015).

advocate? Second, who are the main actors and supporting characters in these stories? Third, what are the prominent discursive themes? Fourth, what are the salient silences and gaps in the narrative? The rest of this chapter evaluates the Long-Term Strategies submitted to the UNFCCC by Ethiopia, The Gambia, Nigeria, South Africa, and Zimbabwe through these four questions.

These five are the first African countries to submit long-term strategies to the UNFCCC in English, as Benin and Morocco submitted theirs in French. Between them they account for over 40% of African GHG emissions and almost one-third of Africa's population. They are significant regional leaders in terms of the African Common Position in the UNFCCC negotiations and their strategies are likely to be influential in shaping long-term climate strategies in the rest of the continent. In the challenges they collectively and variously narrate—including oil and coal dependency, agrarian transformation, external donor dependency, and vulnerability to various dimensions of climate change—they also encompass many of the issues faced by other African countries.[32] Their strategies have many similarities, although they deal with different contexts and challenges. The next section considers their overall narrative plots and the main policies they advocate.

Plots and Policies

The Ethiopian long-term low emissions development strategy aims to reach net zero emissions by 2050 and is closely aligned to the national development strategy, which aims to reach lower-middle-income status by 2025 and make Ethiopia an 'African Beacon of Prosperity'.[33] In this respect, Ethiopia's climate strategy is a key element of a broader aspirational narrative of Ethiopian modernization, built on economic growth rates of nearly 10% per year since 2004, some of the highest rates in the world. Despite this growth, driven by manufacturing and industrialization, Ethiopia is still a largely agricultural economy, and agriculture contributed about half of all GHG emissions in 2020.[34] Politically, the climate strategy has emerged from a period of upheaval: the government has been engaged in violent internal conflict against militia groups from the regions of Tigray, Amhara, and Oromia since 2020. Prime Minister Abiy Ahmed, in power since 2018, was awarded the Nobel

[32] Baek et al. (2019); Müller et al. (2020); Newell and Bulkeley (2017); Newell and Phillips (2016); Power et al. (2016); Simmet (2018).
[33] Federal Democratic Republic of Ethiopia (2023, p. 21).
[34] Federal Democratic Republic of Ethiopia (2023, p. 20).

Peace Prize in 2019 but has since been accused of denying basic rights and committing human rights violations, including 'a largely invisible campaign of ethnic cleansing' in the north.[35] Abiy Ahmed has also overseen 'the beginning of the end of the EPRDF's "developmental state" project, as the new regime adopted a more market-based development strategy with strong support from western donors'.[36] Compared to previous Ethiopian climate policies, the long-term low emissions development strategy reflects this 'shift toward a more private-sector-driven, competitive, and resource-efficient economy'.[37]

The Ethiopian strategy profiles three decarbonization scenarios offering a range of possible trajectories to make the economy more innovative, sustainable and resilient. The LTS is envisaged as contributing to GDP growth and employment, generating 85 million new jobs by 2050, as well as climate resilience. A key policy is the Women and Youth in Green Jobs Program, intended to address 'barriers to accessing economic opportunities'.[38] After farming, deforestation, and land degradation account for a large proportion of Ethiopia's GHG, with the energy, transport, industry, and waste sectors together only contributing about 8% of total emissions. For this reason, the LTS focuses on goals like modernizing the agricultural sector, as well as reducing deforestation, carbon sequestration by afforestation, increasing wildlife habitats, renewable energy and electrification, increased train and urban mass transit, increased manufacturing, and better waste removal. Ethiopia already relies upon hydropower for much of its electricity and in 2040 they forecast electricity will be generated by hydropower (55%) and solar PV, geothermal, and other renewable sources (45%).[39]

The Gambia is another economy dominated by agriculture and developmental challenges, although on a very different scale and political-economic context to Ethiopia. It is the smallest country in mainland Africa and is surrounded by Senegal. Its Long-Term Climate-Neutral Development Strategy 2050 is framed as part of a wider development strategy aiming to become a climate resilient, middle-income country by 2050. Like Ethiopia, its climate strategy has also emerged from a turbulent political context. The Gambia was ruled for over two decades by President Yahya Jammeh, who repeatedly violated political rights and civil liberties. The 2016 election resulted in a surprise victory for opposition candidate Adama Barrow, who was re-elected in 2021.

[35] Callamard and Roth (2022).
[36] Lavers (2024, p. 18). See also Chinigò (2022).
[37] Federal Democratic Republic of Ethiopia (2023, p. 19). In contrast, see Okereke et al. (2019).
[38] Federal Democratic Republic of Ethiopia (2023, p. 11).
[39] Federal Democratic Republic of Ethiopia (2023, pp. 20, 24–7).

Respect for fundamental freedoms including the rights to free assembly, association, and expression have improved under the Barrow administration, but it has faced criticism for continued corruption.[40]

The Gambian economy is based largely on services, agriculture, and tourism, and future electricity needs are predicted to be met by hydropower (56%) and wind/solar (44%).[41] Emissions in the transport sector are forecast to be offset by carbon sequestration through forests, enabling The Gambia to reach net zero by 2050. Much of the strategy focuses on forests and agriculture, including proposals for the implementation of agro-ecology, forest protection and monitoring, afforestation, conservation agriculture, and maintaining 30% of the country's total land area under forest cover. The conclusion is direct and simple: in a nutshell, it argues, this strategy comes down to 'Use of Renewable Energy and Growing Trees'.[42] In language that is distinctively different to the other strategies in its ambition and eloquence, the strategy emphasizes their 'solemn duty and obligation to bequeath a planet in a better state than it is today, for generations yet unborn'.[43]

The Nigerian climate strategy addresses challenges on a completely different scale. Nigeria has one of Africa's largest economies, the largest population, and is the continent's largest oil exporter. In this context, decarbonizing the economy while addressing significant political and socio-economic challenges requires some leaps of the imagination as well as considerable political will. Setting out their vision for tackling these problems, the Nigerian low carbon narrative is one of growth and innovation, transforming a vulnerable, oil-dependent economy to net zero by 2060 through reliance on natural gas as a transition fuel. Oil exports generate 95% of all foreign exchange earnings in Nigeria and contribute 65% of the annual federal budget.[44] Despite the abundance of oil and gas, access to electricity is poorly distributed and often unreliable, meaning many private households and firms depend on self-generated electricity—typically diesel generators—which leads to high GHG emissions and other pollutants.[45] For these reasons, Nigeria is often described as a classic petrostate with an often-dysfunctional policy regime and few political constituencies in support of climate action.[46] However, over the last decade policy discourse in Nigeria has started to change, particularly under

[40] Gardner et al. (2024); Hultin and Sommerfelt (2020); Keita (2023); McConnell and Jallow (2022); Sanneh (2022).

[41] Government of The Gambia (2022, p. 4).

[42] Government of The Gambia (2022, pp. 43, 67).

[43] Government of The Gambia (2022, p. 3).

[44] Federal Ministry of Environment (2021, p. 26). See also National Climate Change Council (2023).

[45] Federal Ministry of Environment (2021, p. 8). See also Elum and Mjimba (2020); Nwozor et al. (2021); Osunmuyiwa et al. (2018).

[46] Macdonald (2017); Osunmuyiwa and Kalfagianni (2017); Van de Graaf (2022, p. 438).

President Muhammadu Buhari after 2015, to the point where Nigeria now has tried to claim a leadership role in terms of climate ambition.[47]

The Nigerian climate vision is that '[b]y 2060, Nigeria will be a country of net-zero emissions across all sectors of its development and climate-resilient with high-growth circular economy in a gender-responsive manner'.[48] The vision narrates a predicted energy transition, calling for the 'development of low-carbon technologies such as carbon capture and storage (CCS), energy efficiency technologies and alternative energy', as well as a range of 'fiscal incentives to promote renewable energy utilisation' aiming at 'increasing the contribution of renewables to the country's energy mix by 50%' by 2050.[49] The expanded strategy published in 2023 relies heavily on nuclear power to meet its net zero targets.[50] There are also targets on increasing access to electricity, reducing deforestation; reducing the carbon footprint of cities by 50% by 2050; removing the politically sensitive gasoline subsidy; expanding rail infrastructure and bus rapid transit systems; moving to a carbon-neutral waste management system at the end of the century; and oil spill clean-up and land remediation in the Niger Delta.[51] A key focus of Nigeria's strategy is a shift from oil to gas, which is positioned as a 'transition fuel' with targets on ending gas flaring by 2030 and increasing domestic gas-powered cookstoves. The vision concedes that 'Nigeria in the foreseeable future will continue to depend on oil and gas for its economic development'. Instead, low carbon development is an opportunity to build resilience and drive diversified growth, creating new 'green jobs' and opportunities in the context of the long-term decline of oil production and rents in Nigeria and the decision by multinational corporations such as Shell to divest their onshore oil portfolio.[52]

South Africa is another large, fossil fuel-producing economy, and together with Nigeria they are the two largest emitters of GHGs and largest economies in sub-Saharan Africa. The South African climate strategy advances a narrative in which the state and policy planners transform a coal-dependent economy through a Just Transition to an envisioned low carbon, net zero economy by 2050. The South African economy is one of the most coal-dependent in the world, relying on it for more than 90% of electricity generation and it is

[47] Death (2024a); Gardner et al. (2023); Nwozor et al. (2021); Okereke (2021).

[48] National Climate Change Council (2023, p. 15).

[49] Federal Ministry of Environment (2021, p. 21).

[50] National Climate Change Council (2023).

[51] Federal Ministry of Environment (2021); Federal Government of Nigeria (2021, p. 22); National Climate Change Council (2023). See also Okereke (2021).

[52] Federal Ministry of Environment (2021, pp. 7, 20, 26, 31–3); National Climate Change Council (2023).

the fourth-largest coal exporting country in the world and a top-20 global GHG emitter.[53] Its political-economic model has been historically described in terms of the Minerals–Energy Complex (MEC), a regime of accumulation based on low-cost, coal-fired, state-owned electricity production, via the parastatal Eskom, and cheap labour, articulated during the Apartheid era through Afrikaner political power and financial capital.[54] Since the early 2000s, however, South Africa has been regarded as a climate leader and is 'both ambitious and unusual as a developing country'.[55] It has played a key role in international climate summits and has a strong trade union and social movement landscape which has advocated for the importance of a Just Transition to a decarbonized economy, while raising concerns about the unjust implications of current South African energy politics.[56]

The South African government's strategy is one of 'transformational rather than incremental change' across the sectors of energy, industry, agriculture, forestry, and waste, driven by the heroic figures of the state and its planning experts.[57] Policy mechanisms include a carbon tax; carbon budgets; phasing out fossil-fuel subsidies; zero emission building standards by 2030; and afforesting 100,000 hectares of land in the Eastern Cape and KwaZulu-Natal.[58] A central focus of the energy transition is, of course, decommissioning 30% of Eskom's coal-generation capacity by 2030 (rising to nearly 100% by 2050) and its replacement by new hydro, solar, and wind. This builds on investments in renewable energy that have slowly borne fruit since the early 2000s, particularly under the auspices of the contested and stop-start Renewable Independent Power Producer Procurement Programme (REIPPPP). Renewable sources have been rising sharply to contribute 10.5% of the total electricity output in South Africa in 2020.[59] The capacity of the ruling African National Congress (ANC) to deliver on this vision was brought into question when it failed to win a majority for the first time in the post-Apartheid era at national elections in May 2024, and had to go into a government of national unity with a range of more conservative opposition parties.

South Africa's neighbour, Zimbabwe, has a very different international status and reputation and is rarely regarded as a leader on climate change.

[53] Republic of South Africa (2020, p. 9).

[54] Baker et al. (2014, p. 797). See also Baker et al. (2021); Bowman (2020); Hochstetler (2020); Swilling et al. (2016).

[55] Tyler and Hochstetler (2020, p. 199). See also Lachappelle et al. (2017, p. 315); Newell and Bulkeley (2017); Power et al. (2016).

[56] Death (2011); Fakier and Cock (2018); Gardner et al. (2023); Hochstetler (2020); Swilling and Annecke (2012); Trollip (2021).

[57] Republic of South Africa (2020, p. 4).

[58] Republic of South Africa (2020, pp. 15, 23–5, 36, 39–41).

[59] Winkler (2022). See also Baker et al. (2021); Hochstetler (2020, p. 48); Rennkamp (2019); Swilling et al. (2016); Tyler and Hochstetler (2020).

The Zimbabwean Long-term Low Greenhouse Gas Emission Development Strategy focuses on low-emission and climate-resilient economic growth and is framed in relation to the broader development strategy to become an upper-middle income economy by 2030. Yet Zimbabwe's development strategy has been characterized as 'disjointed and with weak evidence and limited social embeddedness', and there are doubts over the government's capacity and determination to deliver on a low carbon transition.[60] The high hopes for democratization and renewal after the military coup that ousted President Robert Mugabe in 2017 have been dimmed, as Mugabe's successor, Emmerson Mnangagwa—known as 'The Crocodile' for his violent reputation as Mugabe's security enforcer, as portrayed in Figure 3—faces 'serious legitimacy challenges' as well as struggles to deliver fiscal stability, poverty alleviation, and good governance.[61]

Zimbabwe's narrative is a green growth strategy, identifying policies that will drive growth and development as well as mitigating GHG emissions. It is explicit that, as a developing country, 'strengthening the national economy

Figure 3 Zapiro, 'The Crocodile'

[60] Chimhowu et al. (2019, p. 83). See also Dzvimbo et al. (2023); Müller et al. (2020).
[61] Helliker and Murisa (2020, p. 7). See also Ndlovu-Gatsheni (2009).

and improving the livelihoods of Zimbabweans is an important priority'.[62] In the energy section, the strategy identifies '21 mitigation measures including large hydropower projects (Batoka and Devils George) accompanied by other renewable energy measures such as the introduction of solar PV at the commercial and residential scale', although these are rather vaguely referred to as the 'assumed expansion of new and as-yet unplanned renewable power generation projects'.[63] Key policies also include increasing the carbon tax, stopping net-deforestation by 2030 and increasing the use of conservation agriculture, as well as electric and hydrogen vehicles, biofuels, and expanding public transit. Yet as one study points out, even where policies look good on paper Zimbabwe suffers from 'implementation paralysis'.[64]

In terms of broad goals and targets the five strategies are quite similar, although the scale of the challenge of decarbonizing the energy sector is vastly different in South Africa and Nigeria compared to the other three. As to be expected from the size of their economies and track record on climate change politics, the strategies of South Africa, Ethiopia, and Nigeria are far more detailed and credible than those of The Gambia or Zimbabwe. The next section evaluates the roles of the leading actors and supporting characters in these strategies, where there are also many similarities.

Leading Actors and Supporting Characters

The five long-term strategies all portray the main agents of change as national governments. These are national strategies, after all, authored by national governments and ministries and unsurprisingly they paint a picture of a heroic, ambitious, central state in a leading role. This is typically of LTS more widely, in a broader context in which state-led development planning is returning to prominence in developmental practice and ideology after decades of market-led approaches under the Washington Consensus.[65] These are all top-down narratives in which decarbonization policy will be delivered to 'the people'.[66]

[62] Government of Zimbabwe (2022, p. 1).
[63] Government of Zimbabwe (2022, pp. ix, 7).
[64] Dzvimbo et al. (2023, p. 189).
[65] Ross and Fransen (2017); Ross et al. (2021). See also Alami et al. (2021); Chimhowu et al. (2019); Müller et al. (2020); Newell (2021, pp. 93, 143–4).
[66] The prominence of top-down strategies is widely noted in the literature. See Dzvimbo et al. (2023); Medhin and Mekonnen (2019); Mohamed and Montmasson-Clair (2019); Olaopa et al. (2018); Sinthumule (2016).

South Africa is one of the best examples of this, where the story relies upon South Africa's performed role as a leading player in international climate diplomacy for over two decades. At its heart, the LTS is a state-led heroic story in which slow and tentative steps to challenge South Africa's coal-based economy are to be dramatically scaled up, driven by 'in-depth sectorial transformation plans'.[67] This reflects long-standing enthusiasm and experience within South African policymaking for scenarios, forecasting, and transformative planning.[68] The South African LTS goes into considerable detail about the sequencing of how, when, and where transitions will take place. The strategy details three phases: 'starting right' (prior to 2021), 'turning the corner' (to 2025), and 'massive rollout' (2025 onwards), and these are each elaborated in terms of specific policy packages and parallel planning processes.[69] Given that renewable energy in South Africa is generated by private firms, it is notable that the LTS devotes only four short sentences to 'the role of the private sector', following four pages on state-centric governance.[70]

The Ethiopian strategy is similar, in that it is also based upon a self-asserted 'position as a "green champion" and a "green model" country in Africa'.[71] It uses the strategy to 'demonstrate Ethiopia's global leadership in climate action' as one of 'the first Least Developed Countries (LDCs) to submit an ambitious NDC to the UNFCCC', and it seeks to 'inspire others to increase their contribution to this collective effort'.[72] While Ethiopia has historically followed a development model based on a centralized, coordinating state, there is a strong emphasis on increasing the role of the market and the private sector in this climate strategy.

The Nigerian state also uses its LTS to try to portray itself as a continental leader in terms of climate ambition and has been hailed as a model for oil-producing states.[73] Whether it will turn out to be a convincing leader, given its continuing reliance on oil production and exports, remains doubtful. The Nigerian strategy envisions a heroic partnership between 'government, business, and economic leaders', drawing upon 'market mechanisms' and social, economic, technological and environmental 'innovation' to deliver a

[67] Republic of South Africa (2020, pp. 4–5, 44).
[68] Chimhowu et al. (2019, p. 82); Eatough (2015); Mohamed and Montmasson-Clair (2019); Tyler and Hochstetler (2020).
[69] Republic of South Africa (2020, pp. xviii, 58–62).
[70] Republic of South Africa (2020, p. 18).
[71] Federal Democratic Republic of Ethiopia (2023, p. 13). See also Medhin and Mekonnen (2019); Müller et al. (2020); Okereke et al. (2019).
[72] Federal Democratic Republic of Ethiopia (2023, pp. 3, 13, 18).
[73] NDC Partnership (2021).

low carbon society by 2050 that moves 'away from oil export dependency'.[74] There is an acknowledgement that fundamental transformation of the economy and society are needed to meet climate goals, but there is confidence that this can be achieved by 'a well-managed transition through effective visioning', led principally by 'strong government commitment and leadership' as well as market mechanisms.[75] As in the South African narrative, and indeed in most other LTS, this is a state-centric story in which the central government and its planners are the central characters. Typical phrasings include: 'Government will inform the people . . . as well as ensuring that . . . Government will sensitize consumers to do their part. . . .' Policy inconsistency, inefficiency, and the need for coordination are mentioned very frequently in the text, with an assumption that if planners and politicians can get the governance right then market mechanisms and innovation will deliver.[76] Indeed, the vision claims that 'all that Nigeria now needs is a further development of the political framework and existing institutions to strategically plan and implement the necessary climate change and adaptation measures within a well-formulated long-term vision of development'. The private sector is envisaged as playing a key role in driving innovation and transition, but even here the LTS is clear that '[p]rivate sector needs government enablement to create new investable opportunities'.[77]

To varying degrees, all these strategies position international financial support as a key supporting character for the transition. In a very direct way, international consultants and development agencies played a key role in the preparation and writing of many of these strategies. The French Development Agency (AFD) assisted with the Ethiopian and Nigerian strategies; the 2050 Pathways Platform supported the Nigerian and the Gambian process; the Global Green Growth Institute supported the Ethiopian strategy; and UNDP, the Russian Federation and GFA Consulting Group assisted with drafting the Zimbabwean strategy.

Even more significantly, international finance is key for the implementation of these strategies. The South African narrative, for example, tries to make a strong case that South Africa is 'attractive as a destination for climate finance' and there is an 'unprecedented opportunity for South Africa to absorb climate finance'.[78] While developed countries and other sources of international finance are explicitly criticized for their lack of climate action and financial

[74] Federal Ministry of Environment (2021, pp. 2, 4, 9, 21, 31–2); National Climate Change Council (2023, p. 131).
[75] Federal Ministry of Environment (2021, pp. 4–5, 21).
[76] Federal Ministry of Environment (2021, pp. 7, 12, 15, 17, 20, 22, 31, 37).
[77] Federal Ministry of Environment (2021, pp. 12, 16).
[78] Republic of South Africa (2020, p. 52).

support so far, they are positioned as potential saviours of an energy transition in the Global South.[79] In this respect the South African LTS appears to have been fairly successful, with the announcement of $8.5 billion for financing South Africa's Just Transition at COP26 in 2021.[80]

The Ethiopian LTS also has an extended discussion of how to finance the strategy, with various suggestions for closing the gap between resources and budgets, including attracting more private finance and creating a better institutional environment for attracting international climate finance. The LTS observes that '[d]espite being a leader in climate change action in Africa, Ethiopia lacks comprehensive capacity development efforts for increasing access to climate finance from multilateral, bilateral, and private sector financing vehicles'.[81] It also suggests exploring 'innovative financing instruments' such as bonds, debt for nature/climate swaps, and payments for ecosystem services as ways of attracting financial resources from the international capital markets.

The Nigerian LTS estimates the total cost of its Energy Transition Plan at $1.9 trillion, roughly half of which will have to come from 'international financial aid'. Hence there is the need for 'global resource mobilisation to successfully implement this Strategy', especially given the 'strong uncertainty' about whether developed countries will meet their promises on climate finance.[82] The Gambian LTS has a shorter discussion of finance compared to the others, although it notes that 'success of the LTS is conditional on the provision of international support' and it is explicit about their complete reliance on donor support for current and future climate policy implementation.[83] The emphasis on international financial support in these strategies reflects a key function of national development plans and strategies in what some are calling a 'new era of state capitalism': 'sending signals to investors' or telling a convincing story that represents countries as attractive to international finance.[84] In this respect, key roles of the state in imagined climate futures are as 'promoter, supervisor, and owner of capital'.[85]

So far, all five strategies have presented pathways to net zero by mid-century, based around transitions to renewable energy, carbon sequestration through

[79] Republic of South Africa (2020, pp. 3, 21, 47–53).
[80] Molelekwa (2022).
[81] Federal Democratic Republic of Ethiopia (2023, pp. 97–8).
[82] National Climate Change Council (2023, pp. 19, 94). See also Federal Ministry of Environment (2021, pp. 12, 37).
[83] Government of The Gambia (2022, p. 57).
[84] Pathak (2017, p. 6). See also Alami et al. (2021); Chimhowu et al. (2019, p. 83); Müller et al. (2020, p. 5); Williams and Waisman (2017, p. 8).
[85] Alami et al. (2021, p. 1295).

forestry and agriculture, and low carbon development of transport, cities, industry, waste sectors, and so on. They have all narrated a state-led vision of transition, supported to varying degrees by international finance. The next section shows similar discourses at work in each strategy, but with greater variety in how they are emphasized and balanced.

Discourses of Vulnerability, Growth, and Justice

All the strategies begin by emphasizing the vulnerability of the country to the impacts of climate change. They each emphasize their status as developing countries for whom a growth and poverty-alleviation focus are paramount, and they all at least mention the importance of a just transition while avoiding the language of reparations. However, the balance between these discourses varies across the strategies.

The opening claims of vulnerability are particularly pronounced in the narratives from The Gambia, Nigeria, and Ethiopia. The Gambian LTS begins by observing that 'The Gambia—particularly its historic capital city Banjul, is identified globally as being one of the countries most vulnerable to the effects of climate change', experiencing increased frequency and intensity of drought, flooding, coastal erosion, windstorms, high temperatures, and intense and erratic rainfalls in recent years.[86] The Nigerian Long Term Vision also begins with the assumption of vulnerability and risk. This is framed both in terms of the direct effects of climate change and the economic impacts of energy transitions in the global economy. It is claimed that Nigeria is 'one of the top-most vulnerable countries in the world' to climate change, and 'global trends' including the fact that 'China, European Union, India, Japan, and the United States have decided to go "net zero"' could have 'devastating impacts on Nigeria's economy' given their reliance on oil exports.[87] The Ethiopian LTS opens with the statement that it is 'a highly vulnerable country' to climate change.[88] It has experienced disastrous droughts and floods in recent years and the '2022 drought was the worst in 40 years, severely affecting 7 million people in southern and eastern parts of the country'.[89] As a result, climate resilience is a central discursive theme in these three strategies.

[86] Government of The Gambia (2022, pp. 3–5).

[87] Federal Ministry of Environment (2021, pp. 1–3). See also Ajibade et al. (2017) and National Climate Change Council (2023, pp. 20, 98–9).

[88] Federal Democratic Republic of Ethiopia (2023, p. 12).

[89] Federal Democratic Republic of Ethiopia (2023, p. 19).

Economic growth is a central discourse in all the strategies. It is most dominant in the Nigerian vision, which begins by asserting that 'the transition to low-emission development is indispensable for achieving sustainable economic growth.'[90] Growth is the dominant discursive frame, mentioned 53 times in the Long-Term Vision, an average of 1.33 times per page. But growth is present in the other strategies too: the South African strategy argues that low carbon development can deliver 'tremendous growth opportunities', and the Zimbabwean strategy proclaims that its mitigation measures can 'provide a significant impulse for economic growth.'[91] For Medhin and Mekonnen, 'Ethiopia's vision is based on the notion that green growth is an opportunity, not a constraint, for low-income countries.'[92]

The discourse of the 'Just Transition' is evident in all strategies to some extent, although it is most prominent in the South African LTS where it is an organizing concept, as a result of many years of work by powerful trade unions and a progressive environmental justice movement.[93] The Just Transition is understood here as recognition that the costs of decarbonization are likely to be unevenly spread, in a society that is already very unequal along class, race, gender, and other lines. The LTS insists that '[t]he South African government is committed to ensuring that the transition is just', meaning that the 'negative impacts are not disproportionally borne by the most vulnerable poor and working-class communities who are simultaneously bearing the brunt of the physical impacts of climate change.'[94] Yet there is little concrete detail on how this will be achieved, and the actual progress of renewable energy generation in South Africa has raised significant concerns about the perpetuation of racial inequality. In terms of global climate justice, the South African strategy also contains one of the most explicit recognitions that developed countries must take the lead in making 'deep emissions reductions' in order to 'transform key aspects of the world economy.'[95] In contrast, there is only one reference to making the transition to a 'low carbon economy a just and inclusive transition for all' in the Zimbabwean strategy, and the Gambian strategy contains no references at all to just transition.[96]

[90] Federal Ministry of Environment (2021, p. ii).
[91] Republic of South Africa (2020, p. 46) and Government of Zimbabwe (2022, p. 1), respectively.
[92] Medhin and Mekonnen (2019, p. 292). See also Death (2015) and Okereke (2024).
[93] Hochstetler (2020); Müller et al. (2020, p. 7); Swilling and Annecke (2012); Swilling et al. (2016); Trollip (2021).
[94] Republic of South Africa (2020, p. 56).
[95] Republic of South Africa (2020, p. 3).
[96] Government of Zimbabwe (2022, p. 3). See also Müller et al. (2020, pp. 3–4).

The Just Transition is prominent in the Ethiopian LTS, framed around creating a 'more inclusive labour market' via an 'economic transformation that leads to the creation of economic opportunities and green jobs'.[97] It also notes the 'challenges workers face in vulnerable employment and enterprises in informal sectors'. The Ethiopian strategy goes into most depth on the structural barriers and inequalities facing women and youth and how these can be addressed through decarbonization. It 'recognises the role that women, youth, and vulnerable communities play in climate action', and seeks to 'maximise opportunities for gender and social inclusion outcomes'. The Ethiopian strategy stands out in terms of the extended discussion of gendered inequalities in labour force participation, and gender gaps in leadership and science and technology-related roles, as well as unpaid care work and other financial and social inequalities related to age and gender. The LTS sets itself the task of 'addressing deeper structural barriers' for women and youth, gendered stereotypes and political under-representation.[98]

Perhaps reflecting the example set by South Africa and Ethiopia, there is a marked shift between the brief passing references to the just transition in the Nigerian vision from 2021 and the elaborated discussion in the 2023 Long-Term Strategy.[99] The Strategy gives greater attention to the need for social justice, especially among affected workers and communities, indigenous peoples, and marginalized populations. Nigerian climate policy has also emphasized the need for 'gender-responsive' development, receiving some international recognition for its attention to gender mainstreaming.[100] Yet there is little depth to the discussion of gender in the Nigerian strategy, beyond the observation of 'pervasive gender inequality gaps skewed against women' and repeated injunctions to 'mainstream gender considerations into all actions'. At one point the strategy even suggests that 'women generally need to gain greater confidence and expertise in business management' in order to deliver a more equitable energy transition.[101] Neither the South African, Zimbabwean, nor the Gambian strategy contains substantive discussion of gendered inequalities, meaning their visions of a just transition are effectively gender-blind. The relatively weakness of gendered considerations in these

[97] Federal Democratic Republic of Ethiopia (2023, p. 11).

[98] Federal Democratic Republic of Ethiopia (2023, pp. 11 & 88–90).

[99] Federal Ministry of Environment (2021, pp. 15, 26, 32); National Climate Change Council (2023, pp. 124–5).

[100] Tobin et al. (2018, p. 14).

[101] Federal Ministry of Environment (2021, p. 29); National Climate Change Council (2023, pp. 123–4).

strategies suggests the need for wider reflection on the gaps and silences across these socio-climatic imaginaries.

Gaps and Silences

Analysis of the narratives presented in these long-term strategies reveals a number of gaps and silences. The most important, for the purposes of the argument in this book, concern the lack of an overarching vision of what a decarbonized economy would look like; little sense of how to get there, in terms of overcoming obstacles and enemies; neglect of wider social engagement and democratic participation; a narrow range of political subjects and agents imagined as driving the transition; and the absence of an ecological perspective going beyond anthropocentrism. Furthermore, and perhaps surprisingly, none of these strategies makes a strong case for climate reparations or payments from the Global North for loss and damage.

First, none of these LTS convincingly imagine what net zero or decarbonization really means for an African economy. Other than climate proofing existing growth models by 2050, what sort of future society do they envision? As Newell emphasizes, '[i]f there is to be an energy transition, let alone a transformation of the energy system, not only how we produce energy but for whom and for what will have to dramatically change'.[102] This entails bigger questions about energy markets, consumption patterns, modes of capital accumulation, and colonial legacies than the LTS appear willing to broach. This lack of compelling detail or grand narrative is perhaps unsurprising for policy documents and vision statements, but given these are long-term strategies that states are encouraged to use to stimulate broader discussions about what sort of societies and economies will exist in 2050 and beyond, they are missed opportunities. Indeed, some aspects of these LTS narratives feel akin to jotted ideas for speculative fiction, for example aspirations for a 'carbon neutral oil and gas sector in Nigeria', or calls to develop and invest in 'the internet of things (IoT) and smart home devices'.[103]

Secondly, none of these strategies devote serious or sustained attention to the obstacles to decarbonization, or what Geels describes as 'active resistance by incumbent actors'.[104] For this reason they barely deserve the title of strategies. The Ethiopian LTS has one of the most detailed discussions of the costs

[102] Newell (2021, p. 64).
[103] Federal Ministry of Environment (2021, pp. 26, 31).
[104] Geels (2014, p. 23). See also Osunmuyiwa et al. (2018).

and benefits of decarbonization, but costs are interpreted here in terms of the price of investments in energy and agriculture infrastructure. There is no sense that there are any constituencies or interest groups that might lose out, and the cost-benefit calculations are structured to make the overwhelming case for transition. Everyone is a winner, in these visions, leading to a bland and depoliticized narrative. Obstacles to decarbonization are barely visible in the South African LTS; a brief reference to '[r]esistance to change' is an exception that confirms the rule.[105] A reference to 'dysfunctional' local government is revealing: this is a top-down, managerialist, technocratic policy vision, which assumes that if governance structures can be made efficient, rational, and functional then the just transition can be achieved.[106] The very brief mentions of the largest fossil fuel emitting corporations and entities, such as Sasol, PetroSA, Sapref, and Transnet are merely descriptive, and others like Exxaro and Anglo American are not mentioned at all. This massively underestimates the continued entrenched power of the Minerals–Energy Complex and its addiction to cheap coal-generated electricity. These socio-economic forces showed their strength in May 2024 when pro-coal parties like the MK and Economic Freedom Fighters were able to deny the ANC an outright majority at the electoral polls for the first time since 1994.

In the Nigerian case there is a similar absence of political economy analysis, in the sense of an assessment of who would gain or lose from a transition. Indeed, the Nigerian approach to transition emphasizes the power of innovation and 'vision'; key documents proclaim that 'Nigeria is a nation of vision' and refer to the importance of 'visioning approaches' and 'visioning work'.[107] The Long-Term Strategy emphasizes the need for 'fresh ideas and a radical new innovative way of thinking and doing things', but concludes with the suggestion that the strategy will be 'easy-to-implement', reflecting the weight placed upon rather vague assertions of cross-cutting innovation to drive the transition.[108] There is no mention of key multinational corporations such as Shell anywhere in the text. Yet any serious strategy for addressing climate change must, surely, reckon head on with the vested interest and entrenched power of the fossil fuel industry and other major multinational corporations opposed to decarbonization. Discussion of net zero strategies must 'raise conflicts and

[105] Republic of South Africa (2020, p. xvii).

[106] Republic of South Africa (2020, p. 46).

[107] Federal Ministry of Environment (2021, p. 13). See also Federal Government of Nigeria (2021, pp. 5–6).

[108] National Climate Change Council (2023, pp. 129, 142). The discussion of the 'Challenges/Risks' to implementation is anodyne and depoliticized. See National Climate Change Council (2023, pp. 140–2).

speak honestly about potential winners and losers of a transformation towards a fossil-free society'.[109]

Furthermore, despite the extended focus on finance in most of these strategies and the central role played by attracting international finance, there is little emphasis on the UNFCCC negotiations on loss and damage or on making the case for reparations from the Global North to the Global South.[110] There is no consideration within the LTS of the conditionalities, complexities, and political economies of climate finance which can constrain the political agency of recipients of loans and grants. A decarbonized global economy will involve a restructuring of the global economy, yet none of these strategies discuss what will be required for a just transition globally.[111] As a result, these strategies are characterized by a depoliticized, technical, consensual, and bureaucratic language typical of policy documents more generally as well as the other LTS submissions to the UNFCCC.

Third, they don't reflect a genuine 'social dialogue' about the means and ends of an energy transition, indicating a lack of wider democratic buy-in and legitimacy. All the strategies make relatively brief references to the processes of consultation that underpinned the visions, but there is neither a very convincing sense of the importance of societal participation nor any reflection on the wider challenges to democracy posed by climate change and decarbonization over the mid- or long-term future.[112] Policies to stimulate public behaviour-change and awareness-raising—upon which environmental policy has historically relied without compelling evidence of success—are evident in several of the strategies. The Gambian LTS, for example, recommends the promotion of behavioural measures 'such as sustainable food production, vegan diet and lifestyle, environment and climate friendly consumer practices, primary waste selection, local waste composting, promotion of local and low carbon products, etc'.[113] Reducing democratic politics to stakeholder engagement workshops and awareness campaigns is typical of the top-down, expert, and bureaucratic-led nature of environmental policy planning in much of Africa, as elsewhere in the world. This sits in tension with an emphasis on the importance of wider deliberative processes and forms of democratic participation in literatures on futures planning and long-term visioning, where

[109] Marquardt and Nasiritousi (2022, p. 638).

[110] There is a brief discussion of Loss and Damage in the Nigerian Strategy, see National Climate Change Council (2023, p. 109).

[111] Newell (2019, p. 30; 2021, pp. 135–6); Okereke (2024); Paterson (2021a).

[112] Mohamed and Montmasson-Clair (2019, p. 91). See also Machin (2022); Paterson et al. (2022); Pereira et al. (2018).

[113] Government of The Gambia (2022, p. 61).

strategies are regarded as 'an occasion to engage the general public in the debate around a desirable future'.[114] The Nigerian strategy recognizes that 'culture is a fundamental part of the lives of the population', and suggests that incorporating cultural traditions into climate strategies could help with achieving 'a sense of belonging' and securing greater buy-in for mitigation and adaptation policies, but offers few suggestions for what this might mean in practice.[115] Some of the strategies discuss measures for wider public engagement and debate, but none go any further than the brief suggestions in the Gambian strategy for communication and outreach activities, including 'group and community meetings, partnership sessions, school activities, traditional media, music, song and dance, road shows, community drama, soap operas, puppet shows, karaoke songs and contests'.[116]

Fourth, the ways in which these strategies imagine and narrate political subjects—citizens, workers, consumers, interest groups, activists—are quite constrained. Typically, these strategies are 'devoid of people'.[117] Any low carbon transition will impact real people and communities, whose labour and livelihoods will be transformed as a result of decarbonization policies (or their failure). In the South African climate strategy, for example, despite the repeated assertions of a Just Transition and the experience of collective social mobilization against Apartheid, poor and working-class communities are primarily positioned as 'vulnerable' victims rather than political agents.[118] There are only two brief references to class and to labour as political categories. In the Nigerian vision and strategy, similarly, the references to civil society actors are brief and tokenistic. There are a few passing references to citizens, workers, and communities; the mention of 'lessons learnt from farming and fishing communities' is an exception that illustrates the state-centric and top-down tone of the rest of the narrative.[119] There are some brief references to indigenous communities in the Ethiopian LTS but none of these strategies imagine workers, unions, communities, activists, or movements as political agents.[120] Furthermore, discussion of race, religion, ethnicity, sexuality, age, and disability is almost completely absent and there is no space in these

[114] Pathak (2017, p. 7). See also Adesida (1994); Chimhowu et al. (2019); Newell (2021, pp. 173, 221); Pereira et al. (2019); Ross et al. (2021); UNDP (2017); Urry (2016, p. 138); Williams and Waisman (2017, p. 12).

[115] National Climate Change Council (2023, p. 121).

[116] Government of The Gambia (2022, p. 62). See also McConnell and Jallow (2022).

[117] Miller et al. (2015, p. 66).

[118] Republic of South Africa (2020, pp. 16, 56–7).

[119] Federal Ministry of Environment (2021, p. 20); National Climate Change Council (2023, pp. 121–5).

[120] Federal Democratic Republic of Ethiopia (2023, p. 62).

policy visions for critical discussion of 'hegemonic constructions of masculinity and femininity'.[121] Queer 'foresight research' has drawn attention to how sustainability transitions 'will disproportionately disrupt the lives of sexual and gender minorities', but this is not acknowledged in any of the strategies.[122] Paterson argues that 'imaginaries need to be able to enable people to situate their current lives in their historical contexts in order to facilitate them imagining how those lives may be transformed'.[123] All these climate strategies fall well short of this aspiration.

The final significant blind-spot concerns the environment itself, in the sense of landscapes and biodiversity, ecological relationships, and non-human life-worlds. While all the strategies mention agriculture and forests in varying degrees of detail, these are framed as resources for human development or carbon sinks. For example, the Ethiopian strategy devotes considerable space to discussing agricultural modernization, with various goals for manure management, a shift from cattle to poultry and more 'market-oriented agricultural production'.[124] There is an extended discussion of Ethiopia's forests which 'have among the highest biodiversity in the world and provide a wealth of goods and services resources', and policies for reforestation, afforestation, and carbon sequestration are major elements of the Ethiopian LTS—including a target of 16–20 million hectares for afforestation, for example—continuing a long and politically contentious history of state-led tree-planting projects in the country.[125] The Gambian strategy also devotes many pages to agriculture and forestry, noting the benefits of agro-ecology and calling for 'the urgent need to aggressively grow trees'.[126] The Zimbabwean vision puts a lot of emphasis on conservation agriculture, 'which increases soil organic carbon as well as revenues from farming and livestock management'.[127] In the South African strategy there is surprisingly little on ecological considerations, despite the importance of conservation and biodiversity, landscape and the wilderness, hunting and tourism, and indigenous cultural heritage and concepts like Ubuntu to South African national identity and the economy.[128] The Nigerian Strategy emphasizes the importance of 'Nature-based Solutions', payments for ecosystem services, and accounting of natural capital, but all in the interests

[121] MacGregor (2010, p. 224). See also Newell (2021, p. 9).
[122] Brooks et al. (2022, p. 263).
[123] Paterson (2016, p. 18).
[124] Federal Democratic Republic of Ethiopia (2023, p. 50).
[125] Federal Democratic Republic of Ethiopia (2023, pp. 56, 59). On the history and politics of forestry in Ethiopia, see Death (2015, p. 120) and Maathai (2009, p. 243).
[126] Government of The Gambia (2022, p. 68).
[127] Government of Zimbabwe (2022, p. ix).
[128] Swilling and Annecke (2012, pp. xxi–xxii); Pereira et al. (2019, 2021).

of 'harnessing the power of nature for Nigeria'.[129] None of the strategies discuss indigenous or traditional ecological epistemologies in any depth. Overall, given these are environmental strategies, tasked with envisioning a future socio-climatic landscape, it is striking—although perhaps unsurprising—that they are so anthropocentric in character. There is little recognition of the fact that the climate-changed futures of 2050 will be shared with other species.

The limited research so far on other long-term climate strategies suggests that these features may well be shared more widely beyond Africa.[130] This is problematic if LTS are to play a more significant role in narrating and guiding just transitions to decarbonized futures. Accordingly, we need more research on the role of LTS in political economies of energy transition as well as an expanded research agenda on future narratives of political economic transformation that could inform more genuinely radical ecopolitical imaginaries.

Conclusion

The gaps and silences in these narratives reflect broader tensions and contradictions of environmental politics in an era of state capitalism and climate coloniality. While these limitations are not surprising, genuine political strategies for decarbonization need to be both more imaginatively ambitious and more honest about the balance of forces in the struggle for a just transition. The scope and ambition of the LTS represent a political opportunity to explore more radical and diverse narratives of imagined future climates, and interrogating their strengths, weaknesses, and silences is a crucial task for the political economy of energy transitions.

At this point, sceptical readers may wonder whether if this is asking too much. Is it worth trying to insist that policy documents submitted to the UNFCCC engage more fully with politics, democracy, subjectivity, and ecology? Is it naïve to hope for more radical, urgent, political narratives in such texts? Certainly, many African citizens appear to have lost hope in their political leaders and don't expect them to articulate, let alone deliver, inspiring visions of the future. In Nigeria, the national power company has been variously described as 'Never Expect Power At all' (NEPA), 'Never Expect Power, Please Light your Candle' (NEP PLC), and 'Power Hoarding Company of

[129] National Climate Change Council (2023, pp. 20, 127).
[130] Ross and Fransen (2017); Ross et al. (2021).

Nigeria' or 'Problem Has Changed Name' (PHC of Nigeria).[131] Despite this scepticism towards political leadership and institutional structures, giving up and failing to hold them to account is not an option. Advocates for the LTS urge policymakers to engage with profound questions about the societies, politics, and economies of the future.[132] The existence of these strategies opens a window for activists, artists, and organic intellectuals to demand more radical, just, and diverse visions. There are also examples of other political visions where it is possible to see more imaginative and politicized narratives.

A good example of a more radical policy imaginary with which to close this chapter is the African Union's *Agenda 2063*. It begins: 'We, the people of Africa and her Diaspora . . .' and foregrounds the fight against slavery, colonialism, and Apartheid in the opening paragraph.[133] It sets out a Pan-African vision of a more just future, 'with freedom from fear, disease and want', and calls for present generations to 'act now to shape the future they want'. It imagines 'an integrated, prosperous and peaceful Africa, driven by its own citizens and representing a dynamic force in the international arena'. Pan-Africanism has always had a utopian current, driven by a belief that an alternative future can be built, repairing the humiliations of the past. It envisages, '[i]n short, a different and better Africa'.[134]

This vision encompasses much broader concerns than climate change and the first aspiration is for inclusive growth and sustainable development. It imagines that 'Africa's unique natural endowments, its environment and ecosystems, including its wildlife and wild lands are healthy, valued and protected, with climate resilient economies and communities.'[135] It is a bold statement of modernization and African Renaissance, full of ambitious targets such as the 'hand hoe will be banished by 2025', the hope that 'by 2020 all guns will be silent', and predicting a Pan-African High Speed Train Network and Africa Outer Space Strategy. It calls for the repatriation of 'Africa's stolen culture, heritage and artefacts' and the end of 'all remnants of colonialism' by 2020.[136] If delivering on these worthy goals proved impossible in the stated time-frame, the vision is at least ambitious and is explicitly democratic, anti-colonial, and political.

[131] Obadare (2009, p. 249).

[132] Pathak (2017); Ross and Fransen (2017); Ross et al. (2021); UNDP (2017); Williams and Waisman (2017).

[133] African Union (2015, p. 1).

[134] African Union (2015, pp. 1–2). On Pan-African imaginaries more widely, see Abrahamsen et al. (2023); Ashcroft (2017, p. 109); Bassey (2012); Mbembe (2017, pp. 87–92); Tamale (2020, p. 341).

[135] African Union (2015, p. 3).

[136] African Union (2015, pp. 3–7). See also Ndlovu-Gatsheni (2014); Royo et al. (2022); UNDP (2017).

In further elaboration of the energy transition components of this vision, it envisages that by 2063 'Africa's energy systems will largely be based on renewable energy resources coupled with a strong and to a considerable extent localized manufacturing sector, highly qualified human resources and an integrated energy infrastructure for both centralized (electricity grid, hydrogen systems etc.) and decentralized energy systems (mini- and smart grids, distributed power etc.).'[137] Crucially, this vision is embedded within a more egalitarian and just global economy, where 'there will be a convergence towards sufficiency and responsible well-being where Africa overall increases its energy use, while today's wealthy countries reduce theirs'. This central aspect of a contraction and convergence decarbonization pathway is something that none of the national long-term strategies discuss in any depth. Overall, the Pan-African ecopolitical imaginary can credibly claim to provide 'a more radical vision of modernity' based on 'an endogenous, African based notion of "development" . . . in harmony with nature'.[138]

This text—for all its flaws and limitations—does represent a more politicized vision of the future, one that is more likely to engage citizens in democratic debate about the future they want to see. The style of writing in Agenda 2063 is more grandiloquent and emotive than the technical language of the long-term strategies. It seeks to provide hope and inspiration, and these are emotions that are necessary, alongside those of anger and satirical humour with which this chapter opened.[139] As such, it encourages us to consider other storylines that might encourage richer and more radical visions of African climate futures. For Marquardt and Nasiritousi, the 'challenge going forward appears to be how to open up discussions to alternative imaginaries'.[140] I agree, and the rest of this book takes up this challenge through Africanfuturist climate fiction.

[137] African Energy Commission (2019, p. 11).

[138] African Energy Commission (2019, pp. 11, 13). On critiques of Pan-African imaginaries, see Abrahamsen et al. (2023); Ndlovu-Gatsheni (2014); Royo et al. (2022); Tamale (2020).

[139] For differing perspectives on the emotional dimensions of the climate crisis, see Bennett et al. (2016); Cassegård (2024); Moylan (2020); Munshi et al. (2019); Sultana (2022).

[140] Marquardt and Nasiritousi (2022, p. 638).

3
Why Read African Climate Fiction?

We need more radical and imaginative visions of climate futures. We also need more committed political leadership and implementation of already agreed measures on climate mitigation, adaptation, and finance, better institutions and legal frameworks, bigger and more effective social movements and intersectional alliances, and more honest and robust public debate. I say this to make very clear that, of course, stories by themselves—however exciting, insightful, or inspiring—will not be enough to deliver a just climate future. This chapter seeks to explore the specific contribution fictional imaginaries *can make* to climate politics. There is plenty of dismissive scepticism, alleging that rather than really trying to change the world 'we have done nothing loftier than sit on the couch and read a book'.[1] I share Roland Bleiker's concern that it might be thought 'trivial, or perhaps even irresponsible, to explore aesthetic themes at a time when the world is engulfed by war, genocide, terrorism, poverty, climate change and financial turmoil'.[2] Timothy Clark wonders 'how far is much environmental criticism vulnerable to delusions that the sphere of cultural representations has more centrality and power than in fact it has?' Has ecocritical analysis become a 'kind of diversionary side-show' to the destructive status quo?[3]

This chapter starts by unpacking and developing the case for why we should read climate fiction as a form of climate theory. What do stories *do*, politically? Complex issues like climate change, war, poverty, and pandemics require stories to make sense of them, and the stories we tell have political implications. In this book I seek to argue that taking them seriously and reading them critically is politically important.

By way of an introductory example: in Nnedi Okorafor's novel, *The Book of Phoenix*, the life story of the eponymous (anti-)heroine is narrated orally into a voice recorder, which is discovered centuries later by an old man named Sunuteel in a North African desert. He listens to the story, reflects on how it

[1] Harper (2023).
[2] Bleiker (2009, p. 1).
[3] Clark (2015, p. 21). For other reflections on these questions, see Thaler (2022, p. 2) and Weldes and Rowley (2015, p. 24).

African Climate Futures. Carl Death, Oxford University Press. © Carl Death (2025).
DOI: 10.1093/9780198960775.003.0003

challenges his own understanding of his history and identity, and decides to re-tell the story in his own way, writing down what will eventually become known as the Great Book.[4] This Great Book is the source of much of the genocidal and patriarchal violence of the sequel, *Who Fears Death*.[5] Sunuteel recalls an essay by Roland Barthes on the death of the author and uses this to justify his abbreviations, reinterpretations, and omissions in his re-telling of the story of Phoenix.[6] In the novel's epilogue, another character, Sola—a mysterious white sorcerer who appears in several of Okorafor's stories—scorns Sunuteel for his naivety. He declares that neither Sunuteel nor Barthes took 'into consideration that the story could be and very often was *shaman*, and that the teller was more often than not *medium*. There are ghosts in the machine and spirits in the books, oral or written.' Rather than trying to find out the truth, Sunuteel 'chose to write fiction', and what he wrote 'deformed the lives of many'.[7] The message of this fictional story is that stories can have power.

It is not hard to think of other stories that have had political consequences. Chinua Achebe's novel *Things Fall Apart* is widely regarded as a landmark text that shaped attitudes towards African culture, identity, and the impact of British colonialism around the world.[8] The South African Apartheid state banned a range of novels for their allegedly subversive content, including work by Nadine Gordimer, Es'kia Mphahlele, Miriam Tlali, and Anthony Burgess. Ngũgĩ wa Thiong'o was imprisoned after the publication of his novel, *Petals of Blood*, by the Moi regime in Kenya.[9] The story of Noah and Ham in the Book of Genesis was invoked in the propaganda against the Tutsi that preceded the Rwandan Genocide in April 1994.[10] Protestors against Donald Trump in 2020 dressed as handmaids from Margaret Atwood's novel, *The Handmaid's Tale*.[11] The list could go on. Fictional stories have power and this chapter draws on perspectives from postcolonial, feminist, and queer theory and African-futurism to try to understand their role in contemporary climate politics. As Senegalese novelist Mariama Bâ once argued, '[b]ooks are a weapon, a peaceful weapon perhaps, but they are a weapon'.[12]

[4] Okorafor (2015a).
[5] Okorafor (2018). *Who Fears Death* was originally published in 2010, before *The Book of Phoenix*. Page numbers in this book, however, are from the 2018 edition.
[6] Okorafor (2015a, pp. 227, 231).
[7] Okorafor (2015a, pp. 231–2).
[8] Achebe (1958).
[9] wa Thiong'o (1977).
[10] Hintjens (1999, p. 252).
[11] Atwood (1985).
[12] Quoted in Nfah-Abbenyi (1997, p. 13).

I argue that fictional futures can be read for their contributions to climate theory. Their role is not to communicate climate science more widely or introduce new technological 'solutions', but rather they can strengthen our potential to see the world—ourselves and others—from alternative perspectives. Fiction—especially in the genres of speculative, sci-fi and utopian fiction—has often been associated with radical traditions of politics. As Ursula Le Guin puts it, fiction has 'the power to show that the way things are is not permanent, not universal, not necessary'.[13]

The previous chapter discussed how national climate policies, in the form of long-term strategies, reflect and narrate ecopolitical imaginaries. This chapter argues that novels, short stories, and films also challenge and construct ecopolitical imaginaries. As such it draws upon and endorses Roland Bleiker's passionate defence of aesthetic approaches to world politics:

> Yes, we should embrace aesthetics as an essential aspect of understanding world politics; yes, questions of taste are central to who we are and what we do; and, yes, we can—and indeed should—indulge in artistic endeavours. During destitute times we might, in fact, need aesthetic inspiration more than ever to find innovative solutions to entrenched conflicts and difficult political challenges.[14]

In the discussion below I use another novel by Nnedi Okorafor, *Lagoon*, as an example to illustrate how I find aesthetic inspiration and draw political implications from works of fiction.

After making the case for the political implications of reading fiction, this chapter provides a brief introduction to the burgeoning field of climate fiction more generally, before exploring the role of stories from and about Africa in this field. In contrast to the prominence of African stories and authors in world literature more generally, Africa remains relatively marginal in science fiction and climate fiction. Yet, there is a growing body of Africanfuturist climate fiction and these stories have an important role to play in producing African and global ecopolitical imaginaries. The final section of the chapter introduces the authors and stories that the rest of the book discusses in more depth.

Informed by Africanfuturist contributions to feminist, queer, and postcolonial theory, this chapter develops a conceptual approach for the discussion of the stories in the rest of the book. Subsequent chapters are organized in terms of how the stories challenge the places, temporalities, ecologies, and politics of

[13] Le Guin (2016, p. 375).
[14] Bleiker (2009, p. 1).

dominant ecopolitical imaginaries. This entails asking questions about where they are set, when the story takes place, who is the story about (including other nonhuman species), and what are their political implications. First, however, I try to explain why I think we should read fictional stories for their political implications.

What Do Stories *Do*, Politically?

Stories are everywhere, including in politics, science, economics, advertising, and self-help guides. We are a storytelling species, by which I mean that we seem to be hard-wired to try to impose some kind of meaning, plot, or narrative onto the world we perceive through our senses, and we rely upon and reproduce these stories when we communicate with others. Political reality is never accessed directly or objectively, rather it only 'comes into being only through the process of representation'.[15] As such, language has a 'basic importance', as Frantz Fanon put it.[16] There are many forms of representation but here I am concerned with particular types of stories: those commodified and published as fictional novels, short stories, and films.

These formats tell stories in different ways and to different audiences, but before getting into these differences it is worth reflecting on the political implications of fictional stories more generally. Rather than the aesthetic qualities of these texts in themselves, which is often the focus of literary criticism, I focus on the political implications of reading these texts. This is because, as Bleiker argues, '[r]epresentation is always an act of power', in that it makes some things visible and obscures others.[17] In the process of this discussion, I draw extensively on research in literary criticism—and there are many interdisciplinary connections and resonances between critical approaches to international politics and literary criticism, especially ecocriticism—but this book is not a work of literary criticism.[18] This is not to argue that the literary qualities of fictional stories are unimportant, but instead I have chosen to focus on exploring what stories *do*, politically. I begin this discussion by looking at how stories can help us understand and explain our pasts, presents, and futures.

[15] Bleiker (2001, p. 512). On representation, reality and power, see also Abrahamsen (2003); Edkins (2013); Haraway (2016); Weldes and Rowley (2015).

[16] Fanon (2008, p. 8).

[17] Bleiker (2001, p. 515).

[18] For one take on the difference between literary criticism and political analysis, see Bleiker (2009, pp. 173–6).

Fictional stories—not in themselves, autonomously, but through the ways in which we read, discuss, and interpret them—are important means of making sense of the past and narrate various histories about how 'we' got 'here'. This 'we' is important: whoever 'we' think 'we' are, stories can help us understand perspectives on class, race, gender, sexuality, and intersectionality beyond our own experiences and worldviews. For Amitav Ghosh, while stories can't do everything, they 'can give you a way to think about the world around you . . . to enter other people's consciousness and see the world through their eyes'.[19] This argument has been made most famously by political theorist Martha Nussbaum, who claims that '[n]arrative art has the power to make us see the lives of the different with more than a casual tourist's interest—with involvement and sympathetic understanding, with anger at our society's refusals of visibility'.[20] Political citizenship, she argues, requires empathetic imagination of what it is like to stand in another's shoes, as well as compassion and an awareness of our own vulnerability. These are skills, she believes, that can be learnt from reading fiction.

Fictional stories can illuminate a great deal about power relations in the present. As American novelist and theorist Samuel R. Delany put it, even 'science fiction is not about the future; it uses the future as a narrative convention to present significant distortions of the present'.[21] Reading or listening to and interpreting dominant and insurgent stories offers us a moment to 'to step back, reflect and see political conflict and dilemmas in new ways', even to 'move beyond dogma and promote debate about issues that would otherwise remain silenced or marginalised'.[22] Reading and reflecting upon reading is therefore a way to 'critique conventional ideas' and institutions.[23] Some argue that because fiction tends to be more accessible than other scientific, economic, or political formats, stories can be a way to democratize and increase public participation: almost anyone can engage with a story.[24] Here it is important to understand works of fiction as not simply the products of an individual's creative imagination but as elements of wider social imaginaries, in that they invoke, reproduce, or challenge shared societal worldviews.[25] Novels, short stories, and films are also—in different ways—commodities that are culturally shared and institutionally supported through

[19] Quoted in Brady (2019). On the sense-making power of fiction, see Edkins (2013, p. 282); Ghosh (2021, p. 240); Haiven (2014, pp. 24–5); Hoad (2007); Johns-Putra (2019, pp. 7, 44); Krishnan (2014, p. 145); Said (1994); Seymour (2013, p. 23); Tilly (2002); Yusoff and Gabrys (2011).

[20] Nussbaum (2017, p. 483). See also Steele (2011, pp. 180–1) and Weik von Mossner (2017).

[21] Quoted in Streeby (2018, p. 18).

[22] Bleiker (2009, pp. 2, 11).

[23] Johns-Putra (2019, p. 41).

[24] Raven (2017, p. 165); Zamalin (2019, p. 139).

[25] Moore and Milkoreit (2020).

the industries of publication, distribution, discussion, pedagogy, and criticism, and as such they can illuminate wider attitudes and discourses beyond the text itself.[26]

Moreover, reading fiction—including or especially science fiction—can influence actions in the present that can shape imagined futures. Stories and the socially shared processes of reading and reflecting upon them enable societies to deliberate upon possible futures, fired by the 'impulse to make change imaginable'.[27] Bleiker suggests that they can suggest options and techniques for making political decisions, and Johns-Putra posits that some climate change novels 'enable radical ethical alternatives to emerge'.[28] Ghosh argues that there are some things that *only* fiction can do, such as 'give a voice to the nonhuman . . . give a voice to the things that have no voice'.[29] Some have argued that works of fiction can predict or anticipate technological potential or disasters, working as a kind of assessment of alternative forms of technology and social organization, or can help make for 'better policymaking'.[30] For Bill Ashcroft, postcolonial fiction can 'open up a space for political action that is buoyed up by the possibility, indeed, the probability of social change'.[31] This is quite close to how Kim Stanley Robinson, one of the most prolific and influential contemporary science-fiction writers, describes his political role: he sees it as about 'trying to imagine a plausible history that gets us from here to there', in the 'hope it stimulates discussion and action going forward'.[32] This is mobilized in the service of progressive politics and the radical tradition: as one of his characters declares, '[u]topia is the process of making a better world . . . *Our imagination is stronger than theirs!*'[33]

Utopian politics and praxis have a special place for the role of fictional stories in stimulating the imagination. Thaler sees utopian fictions as working like maps that can help in '[o]rienting oneself in a landscape that is deeply unsettling'. Moreover, he also emphasizes how they can encourage a more hopeful stance and 'supply us with an antidote to the crippling sensation that the "game is over", by demonstrating that things could be otherwise'.[34] For some theorists,

[26] On novels as commodities, see Julien (2006); Krishnan (2014); Nikoleris et al. (2017, p. 309); Steele (2011, p. 178); Trexler (2015, p. 118).

[27] Le Guin (2016, p. 376).

[28] Bleiker (2009, p. 12) and Johns-Putra (2019, p. 54).

[29] Quoted in Brady (2019). See also Ghosh (2021, p. 204).

[30] Harris (2017, p. 182). See also Eschrich and Miller (2018, pp. 19–20); Jasanoff (2015, p. 1); Raven (2017, p. 165); Raven and Stripple (2021, p. 232).

[31] Ashcroft (2017, p. 10). See also Edkins (2013, pp. 285–7).

[32] Robinson (2019, pp. 47, 53).

[33] Quoted in Thaler (2022, pp. 180–1).

[34] Thaler (2022, pp. 24–5, 75). See also Alberro (2024); Ashcroft (2017, p. 5); Crawford (2003); Freedman (2000); Haiven (2014); Imarisha and Brown (2015); Jameson (2007); Moylan (2020); Weik von Mossner (2017, p. 163).

particularly those in traditions of radical Black political theory, writing is a practice of freedom, a form of future-orientated emancipation.[35]

In discussions about art and politics it is not unusual to hear claims about the instrumental function of fictional stories. Stories can be used to communicate a political agenda and persuade consumers, voters, activists, or fighters. Stories are deployed to try to change public behaviour or to shape attitudes to an issue.[36] In terms of climate politics, it is often hoped that stories can personalize and humanize abstract issues of climatic changes and energy transitions, and thereby communicate the urgency of taking action, inspiring more sustainable consumption habits or greater political commitment. Many climate scientists have called for art in general and fiction in particular to help communicate climate change in ways that resonate more deeply than scientific forecasts and scenarios. As Trexler reflects, 'there has been not a little desperation for art to bring home the risks of greenhouse gas emissions'.[37]

Yet, instrumental approaches to fiction—and art more generally—miss some of the most important aspects of an aesthetic approach to world politics. As Bleiker argues, language is more than a means of communication, 'it is the process through which we represent and make sense of ourselves and our surroundings: the cultural crystallization of who we are as people'.[38] Overtly political art—propaganda—can be powerful and may even be needed at the current juncture of climate politics.[39] However, it is not the primary focus here. Rather, I am interested in art that challenges how we think about and represent power, justice, security, nature, and so on. In short: art that challenges our conception of *the political*. Ultimately, the study of the humanities is intended to help us understand what makes us human and how we relate to other people, including nonhumans.[40] Taken together, this is what I have in mind when I argue that fictional texts can be read as a form of political theorizing. Reflecting on fictional stories works to direct our attention to certain issues, places, conflicts, and questions, critique dominant assumptions and help us understand past, present, and future. Speculative fiction can make the familiar appear strange and bring the unfamiliar into clearer focus.[41]

[35] Kelley (2002); Weheliye (2014); Womack (2013); Zamalin (2019).

[36] Braun (2022); Liveley (2022); Moulin (2016, p. 149).

[37] Trexler (2015, p. 75). There are countless examples of this, including Brady et al. (2019); Evans (2015); Hajer and Versteeg (2019); Milkoreit (2016, p. 172); Pereira et al. (2018, 2019); Schneider-Mayerson (2018, p. 474); Wood and Meyer (2022).

[38] Bleiker (2009, p. 4).

[39] On the power of art as propaganda, see Salmon (2010) and Weldes and Rowley (2015, p. 13). On the need for art to communicate climate science, see Klein (2019); McKibben (2019); Oreskes and Conway (2014). For a contemporary novelist who refuses to see art as a policy tool, see VanderMeer (2023).

[40] Bleiker (2009, p. 8). See also Jackson (2020).

[41] Death (2014); Hammond (2021); Melzer (2006, p. 11); Nfah-Abbenyi (1997, p. 20); Rao (2014, p. 202).

It is important not to romanticize fiction as normatively better, or more incisively critical, than other modes of representation. Fiction can obscure certain things or reproduce conservative ideology and racist politics. Climate fiction might misrepresent or purposively distort climate science, promote high consumption lifestyles, legitimize speculative technological solutions or encourage political apathy.[42] Novels and films, in particular, are major industries in which celebrities acquire great fame and wealth whilst the majority of artists—and those who consume these commodities—struggle to make a living. English language novels and films are often powerful ideological tools that have usually worked in the service of imperial and capitalist power relations. As such, we might wonder along with Bleiker whether it is 'possible to speak critically in English . . . to decentre the centre through the language of the centre?'[43] Asking such reflective questions is an essential element of critical analysis. The approach in this book is to explore the ways in which fictional stories both challenge and reinforce dominant power relations.

Novels, short stories and films tell stories in different ways and to different audiences, although all of them can have any of the effects described above. Novels are the preeminent commodity of fictional literature and the staple focus of literary analysis and criticism. They are typically extended narratives that follow a particular character or group over a period of time, providing in-depth understanding of another perspective on the world. Short stories are typically more fragmented and briefer than novels, and are sometimes barely or only partially commodified, often published online and available to read for free. Films are multi-sensory and consumed in a very different manner to text, and range from big-budget, highly commodified productions to free-to-view online short films.[44] Rather than pursue the extensive research comparing and contrasting these formats, in this book I am primarily concerned with unpacking the stories they tell and the similarities and differences in the ecopolitical imaginaries they help construct across these formats.

By way of a brief example, the novel *Lagoon* can be used to illustrate the way in which I seek to read climate fiction as climate theory.[45] This novel tells the story of an alien arrival in contemporary Lagos, and the subsequent interactions between the aliens and an extensive range of human and nonhuman characters. The alien agenda—to bring about social, economic, cultural,

[42] Ghosh (2016, p. 11); Milkoreit (2016, p. 188).

[43] Bleiker (2009, p. 171). See also Howell (2019); Johns-Putra (2019, p. 36); Said (1994); Weldes (2001); Weldes and Rowley (2015, p. 17).

[44] On the different attributes of novels and short stories, see Clark (2015); Johns-Putra (2019); Julien (2006); Kapstein (2016); Rao (2014); Schneider-Mayerson (2018, p. 489). On film analysis, see Joo (2020); Kaplan (2016); Svoboda (2016); Weik von Mossner (2017).

[45] Okorafor (2014).

political, and biophysical change—intersects with some of the desires of the marine life in the coastal waters off Bar Beach, most notably a swordfish whose attack on an oil pipeline opens the book. *Lagoon* is speculative petrofiction which is not explicitly about climate change but which raises issues of anthropogenic ecological harm, resource extraction, adaptation, and sustainability that are very relevant to climate politics. It is not a realistic, predictive, propagandistic, or programmatic story, but it has implications which can be drawn out for climate theorizing, especially in how it narrates processes of cultural awakening, social upheaval, institutional transformation, and political revolution. These implications are developed more fully in the chapters that follow, but four broader insights are worth highlighting here. First, *Lagoon* suggests we should look to African cities like Lagos, so often relatively ignored in global climate change discussions, for visions of climate-changed, Afrocentric global futures. Second, the story implies that restoring better ethical relations with nonhumans should be central to climate politics. Third, I argue it can be read as insisting that embracing human difference—racial, gendered, sexual, differently abled, neuro-diverse—will be vital in coming to terms with changing climates. Fourth, stories like *Lagoon* emphasize the importance of imagining convivial climate-changed futures in which the continuing presence of violence, power, and political struggles is acknowledged.

It is important to read stories such as *Lagoon* critically and interrogate their explicit narratives and implicit assumptions when extrapolating them into wider theoretical claims. In the case of *Lagoon*, for example, in Chapter 7, 'Reparative Climate Politics and Reimagined Subjectivities', I draw attention to the eventual reliance on elite, patriarchal, top-down processes of political transformation, demonstrating the ways in which even revolutionary narratives can be co-opted by potentially quite conservative political imaginaries. Despite this, the novel's subversive central message—that 'even in the corners of palaces, spiders dwell'—is a powerful reminder to be attentive to forms of more-than-human agency and resistance within climate politics and institutions.[46] Perhaps I was also drawn to this tale of beastly revenge against carboniferous capitalism, against those who 'made the world bleed black ooze that left poison rainbows on the water's surface', for more personal reasons.[47] On 24 March 1989, when I was seven, the Exxon Valdez oil supertanker struck Bligh Reef in Prince William Sound—otherwise known as Suungaaciq to the Sugpiaq Native Alaskans—and spilled 10.8 million US gallons of crude oil into the water. World-ending cataclysms exist in the past as well as the future.[48]

[46] Okorafor (2014, p. 259).
[47] Okorafor (2014, p. 3).
[48] Sakakibara (2023, p. 37).

This disaster was a moment of ecopolitical awakening for me, especially as my father worked for Exxon Chemicals. Stories connect with us for very different, sometimes very personal reasons, which is why a reflexive approach is central to reading them politically.

This brief example shares a methodological affinity with an increasing volume of research in International Politics that focuses on political readings of novels, short stories, and films.[49] Some researchers have focused on the circulation of popular culture within social movements or policy-making.[50] Others have sought to analyse the effect of stories on their audiences, either elites or public opinion.[51] Interpretative approaches are more interested in what cultural texts reveals about contemporary discourses and subjectivities.[52] It is the latter approach that I follow here, asking how fictional stories re-inscribe or challenge wider discursive formations that enunciate particular claims about international politics, such as where politics can be found, who can act, how 'we' got 'here' and where are 'we' going, and how to interpret values like justice, peace, sustainability, and so on. For critical theorists, the imagination of social change and alternative worlds is a necessary and constitutive element—although far from sufficient in itself—for the political possibility of radical social transformation.[53]

The genre of science fiction has been particularly attractive to critical theorists, especially in International Politics.[54] This is because of its potential to both reinforce dominant social imaginaries as well as to imagine alternative worlds. In terms of the former, the conservative thrust of much commercially successful science fiction has been noted by many scholars of global politics. For Jessica Langer, the 'figure of the alien—extraterrestrial, technological, human-hybrid or otherwise—and the figure of the far-away planet ripe for the taking, are deep and abiding twin signifiers in science fiction . . . the central myths of the genre'.[55] The masculinist, racist tropes of conquest, genocidal war, and settlement of hostile environments are as deeply embedded in science fiction as in colonial and imperial political imaginaries. Jutta Weldes' studies

[49] Some of the most influential studies for my own approach have included Bleiker (2001, 2009); Edkins (2013); Grayson et al. (2009); Ling (2014); Moulin (2016); Opondo (2016); Rao (2014); Simoneti (2022); Streeby (2018); Weldes (1999, 2001); Weldes and Rowley (2015); Welland (2018).

[50] Haiven and Khasnabish (2014); Imarisha and Brown (2015); Milkoreit (2019).

[51] Carpenter (2016); Daniel and Musgrave (2017); Jones and Paris (2018); Schneider-Mayerson (2018); Young and Carpenter (2018).

[52] Mbembe (2001, 2002, 2017, 2019); Odysseos (2019); Opondo (2016).

[53] Odysseos (2019, p. 347). See also Moulin (2016).

[54] On science fiction and International Politics, see Hunt (2023); Kiersey and Neumann (2015); Simoneti (2022); Weldes (2001). On science fiction and critical theory more generally, see Bould (2009); Freedman (2000); Haraway (2016); Jameson (2007); Melzer (2006); Smith (2012); Streeby (2018).

[55] Langer (2011, p. 3).

of *Star Trek* and Isaac Asimov's *Foundation* series argued that discourses of US foreign policy and globalization were reproduced through science fiction, and thus 'state action is made commonsensical through popular culture'.[56]

On the other hand, many authors and critical theorists—especially feminist, queer, and postcolonial scholars—have emphasized the potential of science fiction to imagine radical alternatives.[57] Techniques of estrangement and defamiliarization—as deployed, for example, in the novels of Octavia E. Butler, Ursula Le Guin, or Samuel R. Delany—can illuminate ways in which contemporary patriarchal social orders, hitherto often taken for granted, can be reinterpreted as fundamentally unjust. Core science-fictional tropes like aliens, cyborgs, and biotechnology populate new universes with 'deviant bodies and subjectivities' that can perform an estranging function for contemporary social orders.[58] For Patricia Melzer, 'some science fiction texts not only incorporate feminist theory but actually produce it'.[59]

This critical function of science fiction is defined by authors like Darko Suvin, Frederic Jameson, and Carl Freedman through the technique or attitude of 'cognitive estrangement'.[60] For Jameson it is the ability of science fiction to 'to defamiliarise and restructure our experience of our own *present*' and perceive the alienating effects of capitalism that render it so useful for critical theorizing.[61] As Freedman explains it, '[t]he science-fictional world is not only one different in time or place from our own, but one whose chief interest is precisely the difference that such difference makes'.[62] Here, the future is not 'a specific chronological register' but is rather 'a locus of radical alterity to the mundane status quo, which is thus estranged and historicized as the concrete past of potential future'.[63]

The definitional debates over the genre of science fiction are intense and extensive, particularly as they relate to cognate genres like fantasy and speculative fiction, often turning on the interpretation of cognition or the prominence of science and technology.[64] These debates are not particularly relevant for this

[56] Weldes (1999, p. 119). See also Weldes (2001).

[57] Influential examples of this, for my own work, include Delany (2009); Freedman (2000, p. 130); Haraway (2016); Hopkinson (2004); Johns-Putra (2019, pp. 115–16); Otu (2021); Seymour (2013); Thaler (2022); Whyte (2018); Zamalin (2019).

[58] Melzer (2006, p. 12).

[59] Melzer (2006, p. 11).

[60] Freedman (2000); Jameson (2007); Suvin (2016). See also Kiersey and Neumann (2015); Miéville (2009); Thaler (2022).

[61] Jameson (2007, p. 286).

[62] Freedman (2000, p. xvi).

[63] Freedman (2000, p. 55).

[64] Examples of these debates include Bould (2009); Delany (2009); Evans (2017); Kiersey and Neumann (2015); Langer (2011, p. 8); Miéville (2009); Raven and Elahi (2015).

book, especially as African science fiction often sidesteps them, as will be seen below. There are further debates over the emerging genre of climate fiction, often argued to be a sub-genre of science fiction, as implied by the label 'cli-fi'. These are more important, so the next section discusses the broader field of climate fiction and the emergence of a canon, before turning to African climate fiction.

Canonical Climate Fiction

Climate fiction resists easy categorization because of the absence of generic styles, conventions, settings, time-frames, and so on. There are some stories in which climate science and/or politics are prominent and others where climate change is merely a background context. There are stories about extreme weather events or disasters, sometimes where climate change is only implied or merely imputed by readers and critics. There are also stories about fossil fuels—often labelled petrofiction—or about wider ecological issues that are hard to read without associating them with climate change but which may not explicitly refer to anthropogenic climate change.[65]

Despite this definitional ambiguity there is a significant body of emerging climate fiction research.[66] There is also an explosion of popular media articles on the phenomenon, with titles like 'The summer that reality caught up with climate fiction', 'How to rethink climate change', and '"Cli-fi" novels humanise the science of climate change'.[67] More sceptical takes include 'What *The Last of Us*, *Snowpiercer* and "climate fiction" get wrong' and 'Climate Fiction Won't Save Us'.[68]

There is a clear canon emerging in the academic and popular references to climate fiction. Examples of twentieth-century climate fiction usually include Ursula Le Guin's short speculative novel, *The Lathe of Heaven*; Arthur Herzog's *Heat*; George Turner's *The Sea and the Summer*; *The Drowned World* by J. G. Ballard; the Earthseed novels by Octavia E. Butler; and *The Ice People*

[65] On petrofiction, see Macdonald (2017); Wenzel (2006).

[66] For an extensive but non-exhaustive list, see Alberro (2024); Chakrabarty (2012); Clark (2010, 2015); Cole (2022); Evans (2017); Fiskio (2012); Gaard (2017, pp. 143–58); Goodbody and Johns-Putra (2019); Johns-Putra (2016, 2019); Joo (2020); Mackey (2018); Mcbride (2023); Mehnert (2016); Milkoreit (2016); Milner and Burgmann (2018); Nikoleris et al. (2017); Rajaniemi (2017); Schneider-Mayerson (2018, 2019); Streeby (2018); Svoboda (2016); Trexler (2015); Urry (2016, p. 43); Wallace-Wells (2019); Weik von Mossner (2017); Whyte (2018); Wood and Meyer (2022); Yusoff and Gabrys (2011).

[67] See Yoder (2023); Clark (2023); and Johns-Putra (2015), respectively. See also Barnett (2023); Brady (2019); Brady et al. (2019); Evans (2015); Pattee (2023); Pierre-Louis (2020).

[68] Harper (2023) and VanderMeer (2023) respectively.

by Maggie Gee. Twenty-first-century examples usually include *The Road* by Cormac McCarthy; Liz Jensen's *The Rapture*; *Solar* by Ian McEwan; *The Broken Earth* trilogy by N. K. Jemisin; the *Science in the Capital* trilogy; *New York 2140* and *Ministry for the Future* by Kim Stanley Robinson; *Flight Behaviour* by Barbara Kingsolver; *The Windup Girl* by Paolo Bacigalupi; the *MaddAddam* trilogy by Margaret Atwood; Saci Lloyd's *The Carbon Diaries 2017*; and many others. There is also an emerging corpus of climate-fiction films, the most famous of which is *The Day After Tomorrow*, as well as several of the Marvel films, *Snowpiercer*, *2012*, *Interstellar*, and *The Beasts of the Southern Wild*. Popular TV series such as *Game of Thrones* have been interpreted as analogies for climate change.[69] While they have attracted much less critical and public attention there is also a burgeoning genre of compendiums of short format climate fiction, including anthologies emerging from Arizona's Center for Science and the Imagination, the three volumes of *Everything Change* and two volumes of *Solar Futures*, and three collections of climate fiction emerging from Grist's short-story competition, *Imagine 2200*, in 2021, 2022, and 2024.[70]

These indicative lists of climate fiction include stories that don't mention climate change at all (*The Road*) and fantasies that bear little relation to climate science (*The Rapture*, the *Broken Earth* trilogy). A narrow definition of 'cli-fi'—a term coined in 2007 by blogger Dan Bloom—as a sub-genre of sci-fi primarily addressing anthropogenic climate change is therefore unsuitable.[71] Instead, I adopt an approach to climate fiction which includes a broad range of stories in which socio-climatic and ecopolitical themes are prominent, and which can be read for potential contributions to climate theory. I am primarily interested in speculative or future-orientated climate fiction, set in worlds that are distinctly different from our own.

In this respect, there are many overlaps between climate fiction and the wider category of environmental literature. Ecocriticism has set out to explore the ways in which nature is represented in fiction and the political implications of these representations. Key themes in climate fiction—and in ecofiction more generally—include responsibilities to future generations, the psychological impacts of living in the Anthropocene, dilemmas raised by technological innovation and scientific debates, and physical survival in hostile environments.[72]

[69] Milkoreit (2019).

[70] E.g. see Eschrich and Miller (2018).

[71] On Dan Bloom, see Johns-Putra (2015) and VanderMeer (2023). In defining climate fiction, I have been influenced by Johns-Putra (2016, p. 267; 2019, p. 62) and Milkoreit (2016).

[72] For some important contributions to ecocriticism, see Azzarello (2016); Clark (2010, 2015); Heise (2008); Huggan and Tiffin (2007); Slaymaker (2001); Weik von Mossner (2017); Wright (2010).

The central question for much of the academic and popular discussion of climate fiction is: 'can cli-fi help solve the climate crisis?' Often, as noted above, there is an implicit or explicit assumption—or hope—that reading climate fiction will motivate people to change their behaviour, reduce their carbon footprint or increase their political commitment to climate-change activism. Ben Okri has called for 'a new art to waken people both to the enormity of what is looming and to the fact that we can still do something about it'.[73] 'Stories move people', declares Robert Moore, Director of the Water & Climate Team at the influential US non-profit advocacy group, the Natural Resources Defense Council.[74] Academic research tends to be more sceptical about how direct or efficacious the link between reading fiction and taking political action might be.[75] More broadly, there is at least a growing consensus that reading climate fiction can diversify and pluralize perspectives on climate politics and potential futures. For Milkoreit, '[t]he diverse storylines of climate fiction make it impossible to think about the future in the singular way'.[76] By reading, watching, or listening to stories about climate change we might develop more empathy for others, as these stories can represent 'human subjects who experience climate change and its political, cultural, social, and psychological implications capture the ambiguities, paradoxes, and uncertainties of human life'.[77]

In this respect, the continued dominance of European and North American storytellers in the climate fiction canon is a concern. Many of the leading researchers in this field have conceded that the mainstream field of climate fiction remains 'culturally narrow'.[78] The genre is dominated by European and North American perspectives, characters, and landscapes, often implicitly projecting a Western viewpoint as a global perspective. Moreover, for Johns-Putra, 'climate change fiction displays a predilection for highly conventional and canonical novelistic techniques grounded in dominant perspectives', and 'highly conventional literary strategies of world-building and character development'.[79] While mainstream climate fiction is

[73] Okri (2023, p. 188).

[74] Brady et al. (2019).

[75] For a selection of research on this question, see Carpenter (2016); Daniel and Musgrave (2017); Jones and Paris (2018); Milkoreit (2017, 2019); Schneider-Mayerson (2018); Weik von Mossner (2017); Young and Carpenter (2018).

[76] Milkoreit (2016, p. 179).

[77] Nikoleris et al. (2019, p. 80). See also Fiskio (2012); Gibson (2020); Pereira et al. (2021); Raven (2017, p. 168); Wood and Meyer (2022).

[78] Nikoleris et al. (2017, p. 310). See also Gaard (2017, p. 145); Johns-Putra (2016, p. 273); Mitchell and Chaudhury (2020); Schneider-Mayerson (2019); Streeby (2018, pp. 4–5); Trexler (2015, pp. 124–5).

[79] Johns-Putra (2018, p. 28; 2019, p. 37).

very good at defamiliarizing some dominant temporal, ecological, and political assumptions—for example, Bacigalupi's portrayal of genetic modifications to food by multinational corporations in *The Windup Girl*; or the interrogation of racism, heteronormativity, and neurodiversity in Octavia Butler's fiction; Margaret Atwood's more-than-human storyworlds; and Kim Stanley Robinson's sympathetic portrayal of collective action and revolutionary politics—the dominant ecopolitical imaginaries in contemporary climate fiction are often set in North America or Europe and plots are typically driven by white men, often espousing liberal and techno-scientific politics. Ghosh sees this as 'a failure of literary forms', which is itself 'an aspect of the broader imaginative and cultural failure that lies at the heart of the climate crisis'.[80]

In the stories listed above, prominent references to African places and people are few and far between. In Ballard's *The Drowned World* and Gee's *The Ice People*, Africa is the goal of the southward migration of the main characters, either for the salvation of a warmer climate to escape the ice or as surrender to the archaeopsychic lure of annihilation in the blazing sun.[81] Kim Stanley Robinson's *Ministry for the Future* portrays a shocking if by now familiar story of cascading climate breakdown, with predictable political upheavals caused by escalating flows of climate refugees from Africa to Europe.[82] In two prominent series of climate-fiction short stories there are more stories set in Africa written by white Americans than stories written by African authors. The three volumes of climate-fiction short stories, *Everything Change*, contain only one story by an African author—'The God of the Sea' by Barakat Akinsiku—as well as another story set in Madagascar, 'Standing Still' by Lindsay Redifer, featuring a white, American teacher telling stories from before the floods to local school children.[83] The *Imagine 2200* short-story collections are explicitly aimed at promoting 'uplifting voices and cultures from the communities most impacted by the climate crisis', and are 'inspired by Afrofuturism, as well as Indigenous, Latinx, Asian, disabled, feminist, and queer futurisms'.[84] The only African stories in these collections are one story by Ebele Mọgọ called 'Omaliyi' set in Nigeria and two stories set in Rwanda and Ethiopia, both by white North American authors.[85]

[80] Ghosh (2016, p. 8).
[81] Gee (1998) and Ballard (1962), respectively. See also Clarke (2013) and Johns-Putra (2019, p. 79).
[82] Robinson (2020).
[83] Akinsiku (2021); Redifer (2016).
[84] https://grist.org/about-imagine-2200-climate-fiction/ (accessed 7 October 2024).
[85] Menzies (2022); Mọgọ (2022); Tanpepper (2021).

In order to enhance the diversity and plurality of landscapes, characters, and perspectives in climate fiction, as well as to offer a fuller consideration of experiences of extractivism, climate apartheid and injustice, conflict, poverty and under-development, racism and intersectional inequalities, the mainstream research on climate fiction would benefit from closer engagement with African stories, settings, and authors. African climate fiction has a great deal to contribute to discussions of global climate politics, as the rest of this book seeks to show. African fiction also has a particular political role and significance that can deepen and extend the debates introduced above, concerning what stories can *do* politically. The role of stories in African political imaginaries and theorizing is discussed in the next section.

African Literature and Modes of Self-writing

In contrast to the relative absence of African stories within mainstream climate fiction, African authors have often been quite prominent in critically acclaimed world literature.[86] African winners of the Nobel Prize for Literature include Wole Soyinka (1986), Naguib Mahfouz (1988), Nadine Gordimer (1991), J. M. Coetzee (2003), Doris Lessing (2007), and Abdulrazak Gurnah (2021). Ben Okri and Damon Galgut are winners of the Booker Prize. Other luminary African authors—including Chinua Achebe, Aimé Césaire, Ngũgĩ wa Thiong'o, Mariama Bâ, Chimamanda Ngozi Adichie, Nuruddin Farah, Ayi Kwei Armah, Amos Tutuola, Tsitsi Dangarembga, Es'kia Mphahlele, Zakes Mda, Buchi Emecheta, and many, many others—have achieved critical and sometimes commercial success in Europe and North America. As a criteria for 'success' this is revealing: Madhu Krishnan has claimed that 'African literature, from a strictly commercial standpoint, is less a cultural artefact written for Africa or an African readership, than one intended for Western eyes or at least predominantly marketed with a Western audience in mind'.[87] For Julien, this phenomenon is referred to as the 'extroverted African novel', whereby canonical literary fictions 'speak outward and represent locality to nonlocal others'.[88]

[86] Ashcroft (2017); Bryce (2019); Iheka (2017); Julien (2006); Krishnan (2014).
[87] Krishnan (2014, p. 36).
[88] Julien (2006, pp. 681–4). See also Eatough (2017, pp. 255–6); Iheka (2019, p. 211); Krishnan (2014).

African literature—and stories more broadly—are thus a central way in which 'Africa' has been imagined, negotiated, produced, and contested in relation to the rest of the world. It is often asserted that African societies are storytelling cultures to a greater degree than societies elsewhere, and oral traditions of myths, stories, and folklore are both important to many Africans and are frequently exoticized and romanticized by Western readers and theorists.[89] Many of the classic, canonical examples of 'African world literature' either take the form of novels in the realist tradition, or integrate folklore, oral traditions, and spirits to produce what has often—and problematically—been described as 'magical realism'.[90] These traditions have had important implications for how African politics, culture, and identity are represented in the rest of the world.

African political theorists—such as Frantz Fanon, Aimé Césaire, and Steve Biko, as well as others writing in the traditions of negritude and radical Black politics—have articulated their own arguments about the critical role of stories, art, and imagination in articulating African identities and negotiating African relations with the rest of the world. For Sylvia Wynter this is one of the foremost African contributions to humanism. She argues that 'the human is *homo narrans*', at least since African ancestors scratched an intricate symbol on a piece of ochre in Blombos Cave in South Africa over 77,000 years ago.[91] Zakiyyah Iman Jackson shows that imaginative work can frequently alter 'the meaning and significance of being (human)'. Exploring the work of authors like Toni Morrison, Nalo Hopkinson, Audre Lorde, Wangechi Mutu, and Octavia Butler, Jackson explores how such texts can 'generate new possibilities for rethinking ontology'.[92] As suggested above, reading and writing novels are themselves a form of theorizing about the world and a way of practising new modes of being and freedom.[93]

African contributions to the project of re-articulating humanism take up the challenge posed by Frantz Fanon, who wrote in the 1950s of the need for a 'new humanism' that would set mankind free and restore human dignity, love, and

[89] Abrahamsen (2003); Ashcroft (2017); Caminero-Santangelo (2014); Gabay (2018); Mbembe (2001, 2006, 2017, 2019).

[90] For a discussion of some of the problems with the genre of 'magical realism' including the difficulty of defining it in ways that aren't just 'speculative fiction by brown people' as alleged by Sofia Samatar, see https://thenewinquiry.com/things-to-come/ (accessed 29 October 2024). For further reflections on oral traditions, folklore and magical realism, see Adejunmobi (2016, p. 267); Aiyetoro and Olaoye (2016, p. 228); Bryce (2019); Hugo (2017, p. 48); Johns-Putra (2018, p. 30); Nyamnjoh (2017); Okorafor (2009); Wenzel (2006).

[91] Wynter and McKittrick (2015, pp. 25, 66).

[92] Jackson (2020, p. 1).

[93] Diagne (2002); Gilroy (2002); Haiven (2014, pp. 228–9); Iheka (2017, pp. 160–1); Kelley (2002); Nfah-Abbenyi (1997); Opondo (2022); Tamale (2020); Weheliye (2014, p. 16).

understanding.[94] For example, Wynter asks 'How do we *be*, in Fanonian terms, hybridly human?'[95] This concern with the politics and possibilities of a new humanism is central to black feminist traditions of thought, which have done most to explore these questions precisely because the dominant imaginary of the human—Man—has always been irreducibly raced and gendered. As Sylvia Wynter puts it:

> This issue is that of the genre of the human, the issue whose target of abolition is the ongoing collective production of our present ethnoclass mode of being human, Man: above all, its overrepresentation of its well-being as that of the human species as a whole, rather than as it is veridically: that of the Western and westernized (or conversely) global middle classes.[96]

For Weheliye, fiction and stories are crucial in order to 'go about thinking and living enfleshment otherwise so as to usher in different genres of the human'.[97] Feminist and ecological stories have been particularly important in this context, and I briefly discuss these below.

African feminists—as well as theorists who have questioned, modified, or rejected the label feminist—have explored the power and potential of fiction in the construction and contestation of identities and subjectivities. For example, the role of motherhood and the family in African literature has received extensive critical and popular scrutiny. The 'Mother Africa' literary and socio-political trope narrates women in African literature as either a 'metonymic figuration of the nation' or as allegories for nature, 'where the land and landscape become identified with the female form and a holistic notion of femininity'.[98] These tropes have been challenged and reworked by feminist authors and theorists drawing on queer and intersectional theory such as Stella Nyanzi, Minna Salami, and Grace Musila.[99]

There are many works of African literature that include a prominent focus on environmental issues, nature or ecology, and there is a growing literature on African ecocriticism. Postcolonial ecocriticism has flourished in Africa with major contributions from Cajetan Iheka and Byron Caminero-Santangelo. Caminero-Santangelo characterizes African environmental fiction

[94] Fanon (2001, pp. 162–3; 2008, p. 1). See also Mbembe (2006).
[95] Fanon (2008, p. 180); Wynter and McKittrick (2015, p. 45).
[96] Wynter (2003, p. 313).
[97] Weheliye (2014, p. 2).
[98] Krishnan (2014, pp. 68, 71). See also Duncan (2005); Jackson (2020); Mama (2001); Newell (1997); Nfah-Abbenyi (1997); Oku (2021); Olaoluwa (2018); Tamale (2020); Wilson-Tagoe (1997); Wright (2010).
[99] Nyanzi (2014); Musila (2022); Salami (2021).

as 'postcolonial regional particularism', which tends to draw 'attention to the ways Africa has been shaped in particular ways as a result of uneven relationships and processes operating at a global scale'.[100] Africanfuturist climate fiction thus is able to draw upon rich traditions of (re)writing and (re)imagining human and nonhuman ontologies in African contexts.

In developing this argument about the specific importance of writing fiction to African political theory, it is worth examining the work of Achille Mbembe in more detail. Mbembe is one of the most influential political theorists to have emphasized the role of aesthetics, storytelling, and fabulation in African political identities.[101] In a seminal article in 2002, he argued that African modes of self-writing struggled to free themselves from the twin pitfalls of anticolonialism and nativism. He argued that while 'contemporary African modes of writing the self are inseparably connected with the problematics of self-constitution and the modern philosophy of the subject', they have been constrained by historicism in the form of Afro-radicalism and nativism.[102] In the first imaginary, 'a manipulation of the rhetoric of autonomy, resistance, and emancipation serves as the sole criterion for determining the legitimacy of an authentic African discourse', whereas the second 'promoted the idea of a unique African identity founded on membership of the black race'.[103] Both these imaginaries lead to political and philosophical dead ends, for Mbembe, in that they are reductive, instrumental, and short-term. He argued, in contrast, for a conception of African identity emerging from historical contingency, the instability of racial categories, and practices of 'selfstyling' even in the midst of conflict.[104] The article generated a wide set of responses, both in the same journal and subsequently, and many other authors have explored the role of African fiction in representing and negotiating the relationship between Africa and the world.[105]

Mbembe has developed and modified this argument in more recent work, exploring ways to problematize the myth of the unity of Black identity and imagine the practices—the 'inventing, telling, repeating and creating variations on the formulas, texts and rituals' of selfhood—of constructing

[100] Caminero-Santangelo (2014, p. 9), and Iheka (2017, 2019, 2021). See also Arseneault (2019); Huggan and Tiffin (2007); Moonsamy (2016); Ogude (2023); cf. Slaymaker (2001).
[101] Mbembe (2001, 2006, 2016, 2017, 2019).
[102] Mbembe (2002, p. 240).
[103] Mbembe (2002, pp. 240–1).
[104] Mbembe (2002, pp. 242, 265, 269).
[105] For immediate responses, see, among others, Diagne (2002); Gilroy (2002); Vergès (2002). For subsequent discussion, see Kotecki (2020); Ndlovu-Gatsheni (2009); Swilling and Annecke (2012). For other perspectives on the relationship between African fiction and political identity, see Abrahamsen (2003); Ashcroft (2017); Iheka (2021); Krishnan (2014); Musila (2022); Tamale (2020).

a 'human community, a same humanity, an essential human resemblance and proximity'.[106] Reflecting on Fanon's 'celebration of the imagination produced by struggle' and the 'resurgence of epic stories', he suggests that the 'Fanonian patient seeks to become, once more, the origin of the future'.[107] He argues that African stories have a distinctive contribution to make to global politics, in particular regarding the ways in which humans are increasingly objectified and objects are increasingly animated. For Mbembe, 'Becoming Black of the World' describes how carboniferous capitalism, itself always predicated on slavery and alienation as forms of power and energy, is profoundly reshaping subjectivity in the global present. By this he means that 'the systematic risks experienced specifically by Black slaves during early capitalism have now become the norm for, or at least the lot of, all of subaltern humanity'.[108] As digital ontologies and shifting ecologies restructure human corporeality in ever more fundamental ways, he suggests that humanity itself is acquiring the subjectivity 'Black' in sense of being 'hard to say which part pertained to the human and which to the animal, object, thing, or commodity'.[109] It is an 'Afrofuturist' aesthetics, for Mbembe, that offers possibilities for reimagining and representing a new form of humanism that goes beyond 'the human' to include assemblages of objects, including the '"non-human", the "more-than-human", the "beyond human", or the "elsewhere-than-human"'.[110] African imaginaries thus have, for Mbembe, a genuinely global significance in terms of the political project of writing new modes of hybrid humanism.

Mbembe's invocation of Afrofuturism is a timely one. Afrofuturism is an artistic and philosophical movement that is directly concerned with understanding the past, present, and future of African identities in a global, postcolonial context, and it has been particularly influential within science fiction. Womack describes it as 'an intersection of imagination, technology, the future and liberation'.[111] Mbembe describes Afrofuturism as 'a literary, aesthetic, and cultural movement that emerged among the diaspora during the second half of the twentieth century. It combines science fiction, reflections on technology in its relations with black cultures, magic realism, and non-European

[106] Mbembe (2017, pp. 28, 173).
[107] Mbembe (2019, p. 141).
[108] Mbembe (2017, pp. 4–6).
[109] Mbembe (2019, p. 8). See also Jackson (2020).
[110] Mbembe (2019, p. 164).
[111] Womack (2013, p. 9). See also Brooks et al. (2022); Dery (1994); Eshun (2003); Iheka (2021); Lavender (2011); Nyawalo (2016); Opondo (2022); Pereira et al. (2021); Rico (2017); Samatar (2017); Streeby (2018); Zamalin (2019).

cosmologies, with the aim of interrogating the past of so-called colored peoples and their condition in the present.'[112] Hitherto, it was often the case that African settings and authors were marginal in mainstream science fiction. Much commercially successful science fiction has been obsessed with futurity and technology, in contrast to racist assumptions of African backwardness and under-development, as well as the fact that many science-fiction stories are overtly militarist and imperialist. For these reasons, African authors and readers often struggled to see a place for themselves and their stories in this genre.[113]

Afrofuturism has been both driven by, and fuelled, a burgeoning market for science fiction in and from Africa, such that some refer to 'a new wave of cultural output that signals the continent as a site from which to imagine the emergence of future worlds.'[114] Mark Bould argues that African sci-fi is 'at least a century old . . . wonderfully diverse and increasingly common.'[115] The generic boundaries between fantasy and science fiction are less clear or useful in an African context because of the so-called magical realist traditions noted above. For this reason I prefer a broader category of speculative fiction—concerned with future or alternative worlds—for the stories collected in this research. Afrofuturist fiction has tended to be animated by anti-imperial, anti-capitalist, utopian imaginaries that challenge racism and seek to envision diverse and plural futures for the continent and diaspora.[116]

As discussed in Chapter 1, 'Introduction', Nnedi Okorafor coined the term 'Africanfuturism' in order to focus on imaginaries that are 'rooted first and foremost in Africa.'[117] Africanfuturist climate fiction opens up some very different concerns and thematics—as well as settings and characters—to mainstream climate fiction as represented by authors like Atwood, McEwan, McCarthy, Kingsolver, and Robinson. In contrast to stories about emission trajectories, the transformation of high consumption societies and Western relationships with nature, Africanfuturist climate fiction insists that the 'founding trauma' of slavery, the plantation and colonialism must be at the heart of the story of the production of contemporary climate change.[118] Carboniferous capitalism has always been predicated on slavery and alienation as a

[112] Mbembe (2019, p. 163).
[113] Dery (1994); Jackson and Moody-Freeman (2009); Okorafor (2017).
[114] Hugo (2017, p. 417). For a small selection of the extensive research on African science fiction, see Adejunmobi (2016); Burnett (2015); Moonsamy (2016); Nyawalo (2016); O'Connell (2016); Omelsky (2014); Otu (2021).
[115] Bould (2015, p. 18).
[116] Gilroy (2002); Samatar (2017); Womack (2013); Zamalin (2019).
[117] Okorafor (2019).
[118] Samatar (2017, p. 180).

form of power and energy. The African experiences of slavery, the Middle Passage, and the colonization of the African continent are now even more crucial elements of a history of modernity in an era of climate change. The Middle Passage—itself famously described as the original alien abduction—is for Mbembe the 'incubator' of modernity and the 'transnationalisation of the Black condition'.[119]

For all these reasons, while African settings and characters have rarely been prominent in mainstream climate fiction, there is great potential for African-futurist fiction to make a significant contribution to wider debates in climate politics. I suggest that African climate fiction can be understood as emerging from much older traditions of African self-writing, speculative fiction, and anticolonial Africanfuturism. The next section, the final part of this chapter, gives an overview of African climate fiction and introduces some of the authors and stories that subsequent chapters explore in greater depth.

African Climate Fiction

Nigeria and South Africa are the powerhouses of African fiction, with the largest domestic markets for literature as well as the most well-known authors, publishers, and literary institutions.[120] The environmental problems associated with heavy industry and the extraction of oil and coal—as discussed in the previous chapter—also means that is unsurprising that much Nigerian and South African literature concerns environmental issues. Writers like Achebe, Habila, Mda, Gordimer, Coetzee, and many others have written powerful fiction on environmental themes. Speculative, future-oriented fiction is also especially popular in Nigeria and South Africa, and there are well-developed science-fiction scenes in both countries. Internationally successful examples include the film *District 9*, co-written and directed by South African Neill Blomkamp, and the *Rosewater* novels by Nigerian-British author Tade Thompson.[121]

The most prominent author writing African climate fiction is Nnedi Okorafor. Based in Arizona (USA), she is a Nigerian-American author of science-fiction and fantasy novels and short stories, for adults and children. Her work has been awarded a number of prizes, including the Wole Soyinka Prize

119 Mbembe (2017, p. 15). See also Dery (1994); Sultana (2022); Whyte (2018); Womack (2013, p. 191); Yusoff (2018).
120 Adejunmobi (2016); Bould (2015); Caminero-Santangelo (2014); Eatough (2017); Krishnan (2014); Wenzel (2006).
121 Blomkamp (2009); Thompson (2016, 2019a, 2019b).

for Literature (2008), the World Fantasy Award for Best Novel (2011), the Kindred Award (2012), the Nebula Award for Best Novella (2016), and the Hugo Award for Best Novella (2016). An adaptation of her novel *Who Fears Death* was reported to be in development for television at HBO. Inspired by authors like Octavia Butler, Virginia Hamilton, Ngugi wa Thiong'o, and Ben Okri, Okorafor's work challenges genres as well as easy categorization. Okorafor's 'postcolonial, post-apocalyptic Africa is a messy, often ambiguous place', as Burnett observes, and her lifeworlds are neither 'traditional' nor 'modern' but are rather intrinsically embedded in what Mbembe terms 'planetary entanglements'.[122]

In the rest of this book I focus on four of Okorafor's novels—*Lagoon*, *Who Fears Death*, *The Book of Phoenix*, and *Noor*—and one short story, 'Spider the Artist'.[123] They are all set largely in West or North Africa and there are extensive intertextual overlaps between them. For example, the plot of 'Spider the Artist'—a romance in the Niger Delta between a woman, Eme, and a spider-droid working as security for an oil pipeline, Udide Okwanka—is referenced in *The Book of Phoenix*, itself a prequel to *Who Fears Death*. These novels tell the story of the collapse of American civilization, in which climate change is accelerated into a fiery apocalypse by a genetically modified organism called Phoenix Okore. *Who Fears Death* picks up the tale centuries later, as a young mixed-race sorcerer, Onyesonwu, goes on a quest to rewrite the Great Book which caricatured the story of Phoenix and legitimated genocidal wars between the Nuru and Okeke populations of North Africa. In contrast, *Lagoon* and *Noor* explore transitions to post-fossil fuel societies in Nigeria. *Lagoon* tells a near-future fable of alien arrival in Lagos, through which the Nigerian petrostate and the more-than-human lifeforms who live in, around, beneath, and outside the city are politically, economically, socially, and ecologically transformed. *Noor* explores the great power but also violence of a wind-powered civilization, in which a technologically augmented mechanic and a Fulani herder are exiled from their communities and come to discover the truth about a vast and destructive sandstorm in the desert.

In the rest of the book, I also discuss stories from five other Nigerian authors. Climate change and nuclear fallout provide the backdrop for Tochi Onyebuchi's novel *War Girls*, in which two sisters struggle to survive and remain human amidst a violent civil war between Biafra and Nigeria, fought with cyborgs and electronic espionage.[124] Terh Agbedeh's short story 'Mango

[122] Burnett (2015, p. 148); Mbembe (2019, p. 93).
[123] Okorafor (2011; 2014; 2015a; 2018; 2021).
[124] Onyebuchi (2019).

Republic' is published in a sci-fi collection entitled *Lagos_2060*, and tells the tale of floods that threaten to overwhelm the model eco-city that Lagos has become, just as the narrator's long-awaited child is born.[125] Osahon Ize-Iyamu's short story, 'More Sea than Tar', is also set in a flooded future Lagos and follows a family struggling to survive amidst the pollution and cut-throat competition.[126] In contrast, Suyi Davies Okungbowa's short story 'Dune Song' is a dry, desert world in which a young girl chafes against the restrictions of her village and longs to explore the winds and sands that cover the long-buried cities of the past.[127] Finally, two short stories by Chinelo Onwualu, 'What the Dead Man Said' and 'Letters to my Mother', are more hopeful tales of societies that have survived and rebuilt after ecological collapse, but in which the trauma and buried violence of the past threatens to erupt.[128]

The most prominent South African author of speculative fiction with climate and environmental themes is Lauren Beukes. Her novel *Zoo City* won the 2011 Arthur C. Clarke Award as well as many other honours.[129] It is a darkly humorous murder mystery set in a near-future Johannesburg. Possibly as a result of climate change or environmental pollution criminals have acquired animal 'familiars' from which they are inextricably attached, on pain of psychic obliteration. Beukes has written several other novels and short stories with cyberpunk themes, as well as journalism and writing for TV. A more recent addition to South African climate fiction is novelist Alastair Mackay, whose first novel *It Doesn't Have to Be This Way* is set in near-future Cape Town as it descends into climate apartheid and drags down the lives of three friends, colleagues, lovers, and activists, Luthando, Malcolm, and Viwe.[130]

I also discuss stories from two other South African writers. Vuyokazi Ngemntu's short story, 'The Serpent's Handmaiden', tells the tale of a messianic delivery of rain in a scorched KwaZulu-Natal, and Jessica Wilson's short story, 'The Drought', like Mackay's novel, starts from the real-life water shortages in Cape Town that led to warnings in 2018 that the city was approaching 'Day Zero' when the taps would be turned off.[131] Wilson imagines how, spurred on by that experience, a future transition to a more ecologically sustainable civilization might unfold.

[125] Agbedeh (2013).
[126] Ize-Iyamu (2019).
[127] Okungbowa (2019).
[128] Onwualu (2019, 2022).
[129] Beukes (2010).
[130] Mackay (2022).
[131] Ngemntu (2022); Wilson (2019).

The most famous writer included here, Doris Lessing, was born to British parents in Iran. She moved to Southern Rhodesia (now Zimbabwe) as a child, where she remained until she moved to London aged 30 in 1949. The influence of growing up in Africa can be seen in many of her short stories and novels and she claimed that 'everything that's made me a writer happened to me growing up in Rhodesia.'[132] The extensive debates about who or what counts as 'African fiction' have already been noted in Chapter 1, and here I support an expansive definition that includes writers who self-identify as African, or who situate their work within traditions of Africanfuturism, or whose work is predominantly set in Africa.[133] According to this definition, Lessing's novels *Mara and Dann*, and the sequel, *The Story of General Dann and Mara's Daughter, Griot and the Snow Dog*, are clearly African fiction.[134] They concern the struggles for survival and ceaseless migrations of the eponymous siblings, across a drought-stricken far-future 'Ifrik' towards an ice-covered and abandoned 'Yerrup'. Although Lessing does not draw on the same styles or traditions of Africanfuturism as many of the other authors discussed in the book there are thematic resonances that merit her inclusion.

There are four more short stories I include in the book, three of them by Motswana author Tlotlo Tsamaase. Her stories 'Virtual Snapshots', 'Eclipse our Sins', and 'Eco-humans' are each disturbing, dystopian visions of technologically and ecologically transformed future societies, in which life is precarious and death comes quickly and painfully.[135] The final short story is by Mame Bougouma Diene, a French/Senegalese/American writer. 'Bittersweet are the Waters' is set in the dried-out basin of what used to be the Atlantic ocean, where insurgents launch a revolt against the private corporation who owns the remaining water.[136]

I also look at three films, all set in a futuristic East Africa. The Marvel films *Black Panther* and *Black Panther: Wakanda Forever* are highly commodified superhero franchises set in the fabled kingdom of Wakanda, an ecologically sustainable, technologically advanced super-state hidden in the heart of Africa.[137] The 21-minute free-to-view sci-fi film, *Pumzi*, by Kenyan filmmaker Wanuri Kahiu, is set in an East African compound, sealed off from a polluted and apparently lifeless desert, and was screened at the 2010 Sundance Film

[132] Bertelsen (1986, p. 134). See also Watkins (2010).
[133] Bould (2015, p. 11); Krishnan (2014, p. 164).
[134] Lessing (1999, 2006).
[135] Tsamaase (2016, 2019, 2022).
[136] Diene (2022).
[137] Coogler (2018, 2022).

Festival.[138] All three films are widely discussed in academic research on Afrofuturism and Africanfuturism and they also raise important and interesting ecopolitical themes that merit inclusion in this book.

This is an extensive list that is broadly representative of much recent African climate fiction, spanning many parts of the continent and some of the diaspora. However, as already discussed, the primary criteria for inclusion is not representativeness, but rather the strength of the challenges offered to dominant ecopolitical imaginaries. These challenges will be discussed in the subsequent chapters in terms of how they imagine the places, temporalities, ecologies, and politics of African climate futures.

Conclusion

The previous chapter concluded on the observation that the official long-term strategies for decarbonization in Africa fail to articulate robust, radical, and imaginative visions of African climate futures. In particular, there are important gaps in terms of the overarching visions of what a decarbonized economy would look like; how to get there, in terms of overcoming obstacles and enemies; the role of wider social engagement and democratic participation; the types of political subjects and agents imagined as driving the transition; and an ecological perspective that goes beyond anthropocentrism. For these reasons, this chapter turned to climate fiction to enrich and deepen our ecopolitical imaginaries.

Reading fiction is clearly not the solution to contemporary climate politics. Indeed, instrumental approaches that treat fiction simply as a resource for more effective communication or stimuli for behaviour change are politically naïve and do not do justice to the most important reasons to value literature. Moreover, fiction can often work to conceal, console, or distract. As Clark suggests, in the face of the debilitating nightmare of the Anthropocene, 'an overinvestment in the power of cultural representations, of the social importance of art and literature, becomes an understandable ethical temptation.'[139] Yet, clearly, the challenges of contemporary climate politics extend beyond the domains of technology, engineering, economics, and governance. Art and literature have important—even crucial—roles to play in how we understand our pasts, diagnose our presents, and open up imaginative possibilities for our

[138] Kahiu (2009).
[139] Clark (2015, p. 196). See also Harper (2023); Heron (2024); Wallace-Wells (2019, p. 144).

futures. Importantly, to reiterate, this is not to claim that these texts in themselves have political consequences, but rather that the ways in which we read, discuss, share, critique, teach, and enact them have political consequences. This chapter has brought together perspectives from postcolonial, feminist, and queer theory to inform an interpretative and critical attitude to reading fiction. Aesthetic approaches to world politics can 'present old dilemmas in a new light' and 'cli-fi at its best lets us travel to climate-changed worlds, to strive there alongside others and then to return armed with that experience'.[140] Crucially, reading, writing, and discussing fiction is an essential element of understanding ourselves and others. This argument has been powerfully articulated by African political theorists such as Achille Mbembe, who emphasize the importance of critically evaluating African self-writing in order to understand contemporary African 'practices of the self', as well as broader African political realities in the context of global capitalism, necropolitics, and climate apartheid.[141] For Beukes, science fiction is politically important 'not for what it predicts about the future of the world, but for how it unpacks who we are in it'.[142]

In this chapter I argued that African climate fiction emerges from much older traditions of African self-writing, speculative fiction, and anticolonial Africanfuturism. There is a growing corpus of African climate fiction although it has yet to receive sustained critical attention in wider research on environmental literature. The next four chapters are structured in terms of the ways in which African climate fiction stories imagine places, temporalities, ecologies, and politics. This structure enables me to both introduce core features of the stories—where they are set, when the story takes place, who is the story about (including other nonhuman species), and what are their political implications—as well as to discuss core aspects of global climate politics. By way of a conclusion, I echo Ben Okri's advice: '[b]eware the stories you read and tell: subtly, at night, beneath the waters of consciousness, they are altering your world'.[143] My world, and my sense of who I am in it, has been transformed by reading these stories. I turn, in the next chapter, to ask 'where in the world might African climate fiction take us?'

[140] Bleiker (2009, p. 178) and Johns-Putra (2015), respectively.
[141] Mbembe (2002, p. 272).
[142] Beukes (2017, p. 330).
[143] Quoted in Okorafor (2015b, p. 26).

4

Strange Places and Unfamiliar Homesteads

One of the great pleasures of reading speculative fiction is trying to work out, in the first few pages, 'where are we?' Writers drop clues about the setting but are rarely explicit: the genre sometimes appears to be full of stories in which working out where and when the story is set, in relation to the writer's *here* and *now*, is one of the principal pay-offs. Nnedi Okorafor's *Who Fears Death* is a good example: the opening chapter offers few clues about the location. It is the funeral of Onyesonwu's father and the 'ceremony was held on the outskirts of town, near the sand dunes'.[1] There are yams, monkeybread fruit, cows, goats and hens. But there are also laptops, water capture stations, laser knives, and solar-powered technology that complicate stereotypical associations between Africa's desert regions and pre-modern poverty. At the end of the first chapter we learn Onyesonwu is telling the story to someone else, somewhere else, four years later, and that she 'grew up far from all the killing. . .'[2] It is not until the last word of the book's epilogue that we are told a more precise, named location. The story finishes in 'the Seven Rivers Kingdom, this place that used to be part of the Kingdom of Sudan'.[3]

A strong sense of place is a key feature of climate fiction. The most common places featured in imagined climate futures tend to be North American and European cities, or coastlines and polar regions on the frontline of global warming.[4] African landscapes are rarely prominent in global imaginaries, except as an apocalyptic warning of dystopian future deserts, slums, and warzones.[5] A dystopian view of African cities was the point of departure for one of the entries to the Post-Fossil City Contest in 2017 by the Urban Futures Studio, entitled 'African alternatives'. The submission juxtaposed the 'status quo' of African cities as 'uncomfortable, unhealthy and often dangerous living environments', with two future alternatives: the 'smart city' and the 'wise city'. The former is 'a vision of the future akin to those crafted by multinational

[1] Okorafor (2018, p. 3).
[2] Okorafor (2018, p. 6).
[3] Okorafor (2018, p. 381).
[4] Luke (2015); Manzo (2010); Mehnert (2016); Nikoleris et al. (2017); Trexler (2015).
[5] Davis (2006); Iheka (2021, p. 10); Myers (2018, p. 468); Pereira et al. (2021).

African Climate Futures. Carl Death, Oxford University Press. © Carl Death (2025).
DOI: 10.1093/9780198960775.003.0004

technology companies and real estate developers', of high security enclaves, robotics and AI, surrounded by nightmarish informal settlements.[6] In contrast, they proposed an alternative future of an egalitarian city with affordable housing, owner operated stores and markets, free public wi-fi, cycle lanes, and indigenous ecosystems. This binary view of African cities as (usually) hell-on-earth or (occasionally) potential paradise evokes much of the wider scholarship on African urbanism.[7] Like Blake Robinson and Karl Schulschenk, the authors of the 'African alternatives' vision, we should be sceptical of elitist smart city visions, but in this chapter I also seek to explore more complex narratives of heterotopian cities and disconcerting living places.

Africanfuturist climate fiction is a powerful tool to illuminate a much greater diversity of places and spaces than is typical within Euro-American climate fiction. Even in these African stories we still get glimpses of ice cliffs calving and a tropical, flooded Manhattan.[8] But the diversity of African landscapes and settings that we can visit through these stories—from markets in Abuja to Cape Town townships, forest compounds and nomadic desert camps, flooded Lagos streets, and the mountains of KwaZulu-Natal—greatly enriches our ability to envision the places of the climate-changed future.

This imagination of distant places—in space and time—was particularly important to me during the years of researching and writing this book. Absorbed with raising a young family in the UK, I read African climate fiction in very familiar, domestic places. Most of these stories I read for the first time at home, in bed, sometimes holding a sleeping infant. I re-read *The Book of Phoenix* in my parents' house, the house I grew up in, watching my children play with their grandparents. I first read *Mara and Dann* in the sunshine during snatched moments of calm shortly after the birth of my second child. I re-read it in similar sunlight, in the same back yard, four years later as our third child turned two. I puzzled out questions and answers while walking the increasingly familiar paths of the Peak District, outside my adopted home of Manchester. Through these stories I encountered worlds that reminded me of places I'd visited and lived in previously, from the deserts of North Africa to plains of East Africa and the mountains and coastlines of South Africa, as well as journeying to countless places that were nothing like anywhere I had ever been. The power of fiction to transport us in space as well as time, and to envision unfamiliar places and other cities and homesteads, is a crucial component of its critical function.

[6] Robinson and Schulschenk (2017). See also Hajer and Versteeg (2019, p. 130).
[7] Iheka (2021, p. 188); Mbembe (2004); Myers (2011); Pieterse (2011); Simone and Pieterse (2017).
[8] Lessing (2006, p. 65) and Okorafor (2015a, p. 8), respectively.

With this in mind, there are three important arguments I want to make that emerge from asking 'where?' questions about African climate fiction. First, I am struck by the unfamiliar, eerie, disorienting locations of many of these stories, in contrast to some classic climate fiction examples that rely on projecting very familiar places into the future. These unsettling landscapes, 'haunted by imagined futures', are a key technique of estrangement through which we can view our own environments more critically.[9]

Second, the portrayal of urban, city locations in many of these stories is interesting not just because they complicate simplistic portrayals of rural, premodern, dystopian African futures, but also because they destabilize utopian and hyper-modernist 'African urban renaissance' storylines. As Hajer and Versteeg proclaim, when imagining 'the transition to a post-fossil future, the city is a crucial site of change', yet 'we see only a few sketches of alternative urban futures'.[10] Furthermore, as Iheka observes, 'the African city has yet to receive the scholarly attention it deserves in the environmental humanities, which tends to focus on rural contexts and spaces of resource extraction on the continent'.[11] Crucially, these stories often focus on the necessity of flight and escape from cities, emphasizing the importance of seeing cities and their hinterlands as spaces of flux.

Third, in contrast to political visions of the future which typically neglect the domestic sphere, climate fictional imaginaries can offer rich, compelling, and diverse visions of future homes, families, and care-giving. Johns-Putra has argued that climate fiction is particularly effective at challenging 'the sentimentality of parental caring, sheltering, and nurturing' and the reproduction of stories about families 'that look like us'.[12] This is true of climate fiction more generally, but in Africanfuturist climate fiction it is striking that many stories portray families and homes as places of danger and insecurity, reinforcing and deepening the arguments of generations of feminist and queer theorists that homes and homelands can be violent places.

There are some powerful political implications to be drawn from all this, implications that can contribute to important discussions about the spatiality of climate change politics and low carbon transitions, and that can help to draw maps that might orientate us better in a climate-changed world.[13] I argue that Africanfuturist climate fiction can help us to demand climate strategies

[9] Gan et al. (2017, p. 2).
[10] Hajer and Versteeg (2019, pp. 123–4).
[11] Iheka (2021, p. 187).
[12] Johns-Putra (2019, pp. 4, 167).
[13] Bridge et al. (2013); Cairns et al. (2022); Newell (2021); Pearse (2021); Swilling and Annecke (2012); Thaler (2022); Tozer and Klenk (2018).

that include people and places more prominently and envision forms of city-hinterland that are fluid, porous, and in-flux, as well as radically transformed domestic environments and care-relations that are imaginative and hospitable.

Unfamiliar Places

Estrangement and defamiliarization are classic features of science fiction.[14] If space and place are at least partially socially produced, then cultural imaginaries can help to view and inhabit places in different ways: to recognize 'the alterability of humanness, space, and place'.[15] There are many techniques by which writers can achieve these effects. Very familiar places can be made strange through a particular shift or innovation; stories can unfold in familiar locations far in the future or unfamiliar locations that resonate somehow with our own context. Environmental and climate activism has often (although of course not always) relied upon very familiar settings—cherished or iconic landscapes, for example—that are threatened or transformed in order to mobilize publics, and this is also true of much eco- and climate fiction.[16] One of the most famous examples is Rachel Carson's *Silent Spring*, whose opening 'A fable for tomorrow' begins: 'There was once a town in the heart of America . . . in the midst of a chequerboard of prosperous farms, with fields of grain and hillsides of orchards', filled with oak, maple, birch, and pine, and clear and cold streams and 'shady pools where trout lay'.[17] It is an explicitly familiar, idealized rural American landscape, unchanged from 'when the first settlers raised their houses, sank their wells, and built their barns'. The 'strange blight', 'evil spell', and 'mysterious maladies' that creep over this familiar landscape are so unsettling because they render a known place strange.[18]

There are many other climate fiction examples that rely upon well-known, familiar settings, albeit transformed or defamiliarized in particular ways. Ian McEwan's *Solar* relies on the stultifying familiarity of suburban England—Reading, London, Oxford—representing the dead weight of the past and a fear of mediocrity or lack of originality that the central character, Michael Beard, finds himself unable to escape even through trips to the Arctic or a solar power plant under construction in New Mexico.[19] Maggie Gee's *The Ice People* is an unsettling and disturbing tale of migration southwards from London at the

[14] Freedman (2000); Jameson (2007); Suvin (2016).
[15] McKittrick (2006, p. 146).
[16] Bridge et al. (2013, p. 335); Death (2016, p. 133); Streeby (2018); Trexler (2015, p. 75).
[17] Carson (1964, p. 1).
[18] Carson (1964, pp. 1–2).
[19] McEwan (2010).

end of the twenty-first century in which much of the estrangement effect is produced by the very different technology, climate, and social norms that have emerged in the otherwise quite familiar landscapes of England, France, and Spain.[20] Megan Hunter's *The End We Start From* is a sparse novel about a displaced family with a new baby who are forced to leave their London apartment by floods and flee to higher ground. Through refugee camps and dangerous towns, they eventually reach relative sanctuary in Scotland, first on the mainland and then an island community.[21] The story ends with their return home to London and the child's first steps in a safe domestic setting. Heteronormative reproductive futurism is reassuringly secured.[22] In Lydia Millet's *A Children's Bible*, a lake-house vacation setting not too far from New York City and the Appalachian Trail is rendered unfamiliar after a climate-charged hurricane and the collapse of parental authority and responsibility, although this story is more ambiguous about the eventual reassertion of familiar domesticity.[23] Most starkly, Cormac McCarthy's *The Road* is set in an almost—but not quite—unrecognizable East Coast, Southern US landscape, where much of the pathos of the story is fuelled by the memories of a happier time for the family, before the unspecified cataclysm. The story's conclusion holds out at least the glimpse of a future within a family group and more hospitable places.[24] Major Hollywood blockbuster films also rely on this trope of very familiar landscapes being transformed by climate change. The iconic scenes of destroyed New York and Los Angeles cityscapes in *The Day after Tomorrow* and *2012* are emblematic in this regard, in that they provide 'now-familiar visuals of what ecological collapse may look like' and portray 'the end of the white world as the end of the world'.[25]

Prominent Afrofuturist films encompass a range of techniques of defamiliarization and estrangement. At one end of the spectrum, *District 9* conjures up familiar urban South African landscapes, including the Johannesburg skyline and township slums, now rendered strange by the alien mothership overhead and the 'prawns' interned in the titular camp.[26] For van Veuren, this film illuminates how 'visceral bodies, institutional bodies, classificatory bodies and bodies of the built environment are placed in crisis by uncertainties around the permeability of their borders'.[27] The *Black Panther* films, in contrast, do

20 Gee (1998).
21 Hunter (2017).
22 Edelman (2004).
23 Millet (2020).
24 McCarthy (2006). See analysis in Johns-Putra (2019).
25 Emmerich (2004, 2009). Quotes from Joo (2020, pp. 73–5). See also Weik von Mossner (2017, p. 156).
26 Blomkamp (2009).
27 van Veuren (2012, p. 582). See also Smith (2012); Veracini (2011).

invoke somewhat stereotypical 'African' rural and forest landscapes, but the twist is that these are explicitly projected as a cover for the hidden kingdom of Wakanda, whose utopian futuristic cityscapes combine traditional aesthetics with hyper-modernist visions of 'smart cities', full of gleaming metal and glass skyscrapers, rooftop gardens, and monorails.[28]

Africanfuturist films that go much further in portraying unfamiliar and disorienting landscapes include *Pumzi*, *Crumbs*, and *Neptune Frost*.[29] *Pumzi* is set inside a futuristic compound, the Maitu community, in East Africa 'after the water wars', surrounded by a desolated, post-apocalyptic landscape. There are no recognizable landmarks in this 'profoundly reconfigured' Africa, and the 'ordered and hierarchical space' of the compound is more akin to a space station than an urban African landscape.[30] It 'runs on electricity produced by grim-faced people on treadmills . . . [and is a] sterilised space, socially and hygienically, that has closed itself off from the outside world'.[31] *Crumbs* is a surreal, dream-like quest through post-apocalyptic Ethiopian landscapes, in which a scavenger hunts through the detritus of late-twentieth-century pop culture beneath a mysterious alien spaceship. *Neptune Frost* is a surrealist musical adventure set in the mine-ravaged landscapes of Burundi, in which people and places are rarely what they seem and cyberspace is as real as the unearthly environments through which the characters pass. These films envision places akin to Francis Nyamnjoh's frontier Africa, where 'everyone and everything is malleable, flexible and blendable, from humans and their anatomies, to animals and plants, gods, ghosts and spirits'.[32]

These unreal, disorienting, unfamiliar places are more characteristic of contemporary Africanfuturist climate fiction, I argue, than the familiar landscapes of much European and North American climate fiction. Of course, there are exceptions on both sides. Margaret Atwood's *MaddAdam* trilogy is set in an unsettling and disorienting landscape, whereas Okorafor's *Lagoon* is intended evoke a Lagos cityscape familiar to contemporary Nigerian readers, despite the disconcerting range of nonhuman and more-than-human occupants and alien visitors.[33] There are more homely settings in short stories by Terh Agbedeh and Chinelo Onwualu that will be discussed later in the chapter.

[28] Coogler (2018, 2022); Zeleza (2018).
[29] Kahiu (2009); Llansó (2015); Williams and Uzeyman (2021).
[30] Omelsky (2014, p. 39) and Iheka (2021, p. 32), respectively.
[31] Omelsky (2014, p. 40).
[32] Nyamnjoh (2017, p. 258).
[33] Moonsamy (2020).

Notwithstanding these exceptions, there is something significant about the prominence of unfamiliar and disconcerting landscapes in Africanfuturist climate fiction. There is of course a risk here that such defamiliarization is exactly what a Western readership is looking for in exoticized 'African' literature, something 'strange and monstrous'.[34] This is a real danger, but it is also important to emphasize that the unfamiliar places imagined in African climate fiction are not typically those of mythic, tribal, primeval wilderness beloved of romantic Africanists, an Africa that is 'incomprehensible, pathological, abnormal'.[35] Rather, these stories are set in apparently recognizable cityscapes, villages, and homesteads; windmills, monorails, and sewers; forests, deserts, and plains. These are African places that could, from one angle, be anywhere in the world. And yet they are reimagined obliquely, rendered disconcertingly eerie, estranged, and defamiliarized.

Okorafor's *Who Fears Death* is a good example. On one level, it is set in the small towns and desert environments of the Sahel. These places, however, are not quite what they seem. As Onyesonwu's training as a sorcerer progresses she becomes more aware of the spirit world, the wilderness, a parallel realm into which magic-uses can pass and travel. She travels to spy on her father, visits forests with her mother, and is visited by mysterious and powerful masquerades. She learns to stand with 'one foot in the spirit world and the other in the physical world . . . part blue energy and part physical body. Half alive and half something else.'[36]

Okorafor's desert landscapes are also more mysterious and magical, more unsettling than they initially appear. There are devastatingly powerful 'Ungwa storms' and the red Vah people live a utopian life of sanctuary and polyamory inside a sandstorm. There are dangerous caves full of ancient technology, juju, hanging corpses, and thousands of white spiders.[37] The desert is a sanctuary from the dangers of the towns and villages. Before she is born Onyesonwu's mother talks to her foetus: '"The human world is harsh", Najeeba said. "But the desert is lovely. Alusi, nmuo, all spirits can live here in peace."'[38] The desert is full of life and Onyesonwu is at home there; this is not a story about thirst, hunger, or oppressive heat. Instead, the desert is a place of beauty, friendship, and travelling companions.[39] Onyesonwu's singing brings desert owls to their camp. Camels play an important role in the narrative, and are

[34] Mbembe (2001, p. 1).
[35] Mbembe (2004, p. 8).
[36] Okorafor (2018, p. 279).
[37] Okorafor (2018, p. 329).
[38] Okorafor (2018, p. 24).
[39] Kotecki (2020, p. 169).

recurring symbols of compassion, intelligence, and bravery.[40] But the desert is also dangerous: Nuru raids seem to appear from the desert and the rape of Onyesonwu's mother, Najeeba, took place there. For this reason, the nomadic Red People, the Vah, who live in the desert, are protected by a magical travelling dust storm. They are 'a nomadic society who hold children in common, practice free love, and value open communication'.[41] 'Living with the Vah people was odd', Onyesonwu reflects, and she questions their norms and rules, especially concerning sex and intimacy. But her experience of their way of life and their spirituality constitutes a crucial transformative experience in Onyesonwu's life.[42] The Vah help Onyesonwu to pass a major test as well as saving her life when she is cursed by her biological father.

The disturbingly unfamiliar desert realms of *Who Fears Death* provide the opening and closing context for the prequel, *The Book of Phoenix*. Whereas most of *The Book of Phoenix* is set in recognizable landscapes of North America and West Africa, the book begins with the elderly Sunuteel travelling in the desert and being forced to take refuge from an Ungwa storm in an ancient cave, full of abandoned computers. One of the computers turns on at his touch and transmits the story of Phoenix Okore to his portable.[43] He reads as the storms rage outside and at one point he sees the figure of long-dead Phoenix herself, dancing on the dunes.[44] Okorafor's novel *Noor* is also set in the Sahelian desert, this time in northern Nigeria, and it is filled with disorienting landscapes, eerie abandoned warehouses, terrifying power structures and mobile, hidden villages and cities. At the heart of the novel is the awesome Red Eye storm that ravages the deserts of northern Nigeria.[45]

Strange desert worlds are recurring settings in Africanfuturist climate fiction. Suyi Davies Okungbowa's short story 'Dune Song' is set in an unrecognizable far-future where sand covers the long-dead cities of the past. This desert is ambiguous and dangerous, and as in Okorafor's *Who Fears Death* the sand dunes exist outside normal time and space. Those who enter them do so to test themselves and acquire self-knowledge. 'Dune Song' seems to be set in a far future West Africa where the chiefs of the settlement of Isiuwa are scoured by 'dust-laden wind' and tell stories of the old worlds before the punishments of water then sand, and there are rumours of civilizations buried beneath the

[40] Okorafor (2018, pp. 218–20).
[41] Kotecki (2020, p. 172).
[42] Okorafor (2018, pp. 270–2).
[43] Okorafor (2015a, p. 4).
[44] Okorafor (2015a, p. 211).
[45] Okorafor (2021).

dunes.[46] The citizens of Isiuwa are forbidden from leaving the settlement and going out into the dunes, kept in by bamboo fences and 'words planted in the mind'. But rebellious subjects, like Nata and her Mam, or the chief's son Tasé and his mother, go out anyway to seek freedom in the sandstorms and whirlwinds. This is not a story about finding a utopian future but of escape from the restrictions of society and the promise of wild, necropolitical otherness. Nata leaves without saying goodbye, hoping 'the sand will flow and sweep over all of Isiuwa like a great ocean', before heading out into the 'roaring ghoul of black wind'. She sees the whirlwind for the first time:

> The Chief is right in calling it the breath of the gods, because it holds within it a crackle, light, and lightning, embraced by wind roiling within itself, gloved in sand and dust and debris. It moves like a cloud would if it were angry. It roars mightily now, up close, as if made of mouth alone. Sand hisses in its wake, an unending flute, an orchestra of whistles, a posse of snakes. Glorious.

It is a whirlwind 'of time, of gods, of opportunity, of liberation', and '[w]herever it would take them, they do not know, but at least one thing is sure: that their tongues and their bodies and their hearts will belong to them'.[47] The desert represents a place of freedom but also social death.

There are many examples of other uncanny, disconcerting, unsettling places in African climate fiction. Doris Lessing's *Mara and Dann* follows a sister and brother as they travel across a far-future Ifrik, through the sun-bleached deserts of what was once Central Africa, up the Cong river to places where forests now cover the Sahara and long-sunken cities can be glimpsed beneath the marshes. Over the Middle Sea the ice cliffs of Yerrup groan and are starting to melt. A chance to study an old and rare map profoundly shakes Mara and her brother as they grapple with the scale of the world and their lack of knowledge about it.[48] Tlotlo Tsamaase's short story 'Eclipse our Sins' is set in the kingdom of Kang in the far-future, where dust storms and a hostile sun keep the terrified inhabitants stuck inside forbidding tower-block cities, with their '[t]eeth of metal and jagged glass. Nothing homelike about this city'.[49] Mame Bougouma Diene's short story 'Bittersweet are the Waters' takes place in the middle of the dried out Atlantic basin, the ocean reduced to a lake. Water has become the currency, status symbol, and object of religious fervour for the surviving

[46] Okungbowa (2019).
[47] Okungbowa (2019).
[48] Lessing (1999). See also Watkins (2012).
[49] Tsamaase (2019).

communities.[50] A flooded world—similarly unsettling and disturbing—is the landscape for Osahon Ize-Iyamu's short story, 'More Sea than Tar', in which mud, rubbish, sewers, flood waters, and human habitations bleed into one another.[51] Lauren Beukes' *Zoo City* takes place in an ostensibly familiar, contemporary Johannesburg, but through the eyes of the central protagonist, Zinzi December, we are drawn into the murky depths of the city's mines, the 'rotten heart of leafy suburbia', the sewers, tunnels, and criminal underworld, where 'the animalled' live in constant fear of being swept away by the inexplicable Undertow.[52] For Lisa Propst, the novel is full of 'conspicuous silences' through which it suggests 'the limits of Beukes's own ability to translate Hillbrow lives into fiction'.[53]

Africanfuturist climate fiction is thus full of strange and unsettling places. Rendering the familiar unfamiliar is a powerful critical technique which can lead us to view our own surroundings and locations—our own environments—from a different perspective. While the wider science fiction and climate fiction genres contain quite a few examples of extra-terrestrial settings in which Earth has been left behind or rendered uninhabitable, there are relatively few examples of this from Africanfuturist authors.[54] The landscapes of most Africanfuturist climate fiction are resolutely earthly, but they are also profoundly unfamiliar.

One way to view these unsettling places is as examples of heterotopias. The concept of the heterotopia, as an 'other space' or 'countersite', was proposed by Michel Foucault.[55] The concept was popularized within science fictional worlds by Delany who titled his novel *Triton: An Ambiguous Heterotopia*.[56] Heterotopia are spaces of 'Otherness, difference and liminality', 'sites of alternate ordering' often associated with landscapes such as 'forests, deserts, jungles, and slums'.[57] These are not utopian spaces of liberation nor dystopian sites of oppression but rather they are 'fundamentally disturbing places' that unsettle the familiar and 'disrupt and test our customary notions of ourselves'.[58] They are not free from hierarchies and power relations, but their

[50] Diene (2022).
[51] Ize-Iyamu (2019).
[52] Beukes (2010, p. 82).
[53] Propst (2017, p. 415). See also Bethlehem (2015); Cloete (2015).
[54] Some of the most well-known of the former include Robinson's Mars trilogy (1992, 1993, 1996). Isolated examples of the latter include Hartmann (2015) and Resnick (1998).
[55] Foucault and Miskowiec (1986).
[56] Delany (1976). See also Alberro (2024, pp. 89–90); Jameson (2007, p. 144); Nyawalo (2016); Zamalin (2019, p. 113).
[57] Hetherington (1997).
[58] Johnson (2006, pp. 84, 87).

order and structures are uncanny, ambiguous, contrapuntal. As well as deserts and forests, urban sites like museums, libraries, gardens, and cemeteries are often described as heterotopian.[59] Later chapters will explore further examples of heterotopian places in African climate fiction, including the museum in *Pumzi* and The Centre in *Mara and Dann*. Through the envisioning of African climate futures set in disconcertingly unfamiliar places these stories lead us to look at our own worlds askance, to question the hitherto familiar order and structures which surrounds us. This technique of estrangement can be politically powerful, as seen in the ways in which African cities are imagined in climate fiction.

African Cities of the Future

Viewing African climate futures from places that are rendered unfamiliar and heterotopian can help to destabilize the powerful discursive binary in which cities in general, and African cities in particular, are imagined as either dystopian or utopian. On the one hand, the European Commission's long-term vision for a climate neutral economy declares that cities are 'the laboratories for transformative and sustainable solutions'.[60] Yet, on the other hand, for many people, including many who would describe themselves as environmentalists, cities are imagined as dirty, crowded, and unnatural. Some advocates of conservation in Africa come close to representing cities as profoundly foreign intrusions on the 'wild continent'.[61] The simplistic—and often racist—imagined associations between Africa, rural landscapes, and pre-modernity have troublingly enduring power. Iconic savannah African landscapes are still often represented as largely uninhabited by people, except indigenous or 'tribal' societies who are close to nature and almost part of the environment.[62] Rather than remaining trapped within the binary of utopian and dystopian imaginaries, African climate fiction can help us envision more complex pictures in which the African urban landscape is 'a heterotopian space that is impure and multiple'.[63]

[59] Foucault and Miskowiec (1986, pp. 25–6); Hetherington (2011); Jameson (2007, p. 99); Nielsen et al. (2021); Nyawalo (2016, p. 219).
[60] European Commission (2018, p. 22).
[61] BBC (2013).
[62] Abrahamsen (2003); Caminero-Santangelo and Myers (2011, p. 15); Iheka (2017, p. 10); Nyamnjoh (2017).
[63] Nielsen et al. (2021, p. 886).

African cities in any form are hard to find in dominant socio-climatic imaginaries. This is of course problematic, especially when some forecast that most of the world's largest cities in 2100 will be in Africa, many with over 40 million inhabitants.[64] African climate fiction is a rich medium for redressing this omission. These stories encompass cities like Abuja, Brazzaville, Cape Town, Johannesburg, Kinshasa, Lagos, Nairobi, Onitsha, and many others unnamed or renamed. As the narrator in 'Virtual Snapshots' dryly remarks, '[o]ur houses have corners, and we don't have lions or animals of the wild parading the city centre'.[65] Several of these Africanfuturist stories focus on Lagos, forecast to become the world's largest city with over 88 million inhabitants by 2100.[66] For example, the novel *Lagoon* is dedicated to 'the diverse and dynamic people of Lagos, Nigeria—animals, plant and spirit', and concludes with its narrator asking 'What is that sweet taste I feel with my feet? It is patriotism, loyalty. Not to the country of Nigeria but to the city of Lagos . . . my glorious city'.[67]

Lagos is a paradigmatic example of the urban–utopia discourse in an African context. Dutch architect Rem Koolhaas famously claimed in 2001 that Lagos was 'at the forefront of globalising modernity' and the rest of the world was merely trying to catch up.[68] Lagos encapsulates the 'Africa Rising' narrative, in which dynamic, youthful, urban populations drive growth, modernization, and sustainability in Africa.[69] African cities like Lagos have become idealized by some commentators as engines of growth, sites of innovation, incubators of middle-class democratic citizenry, and, in the form of futuristic 'smart cities', the key to tackling ecological sustainability and climate change.[70] As Garth Myers reports, 'grand scheme, blueprint master planning' is back in cities like Lagos, Dakar, Nairobi, Lusaka, and Addis Ababa, representing 'the return of the modernist dreams for the continent's cities'.[71] While Myers is critical of the inflated rhetoric surrounding African hyper-urbanism, he still concludes that 'Africa's cities, even in the worst-case scenarios of political violence and economic decimation, are engines of cultural productivity, inventiveness and ingenuity the likes of which the world has rarely seen'.[72]

[64] Satterthwaite (2016).

[65] Tsamaase (2016).

[66] Satterthwaite (2016).

[67] Okorafor (2014, pp. v, 292–3). See also Okorafor (2015b, p. 25).

[68] Quoted in Hugo (2017, pp. 46–7). See also Iheka (2021, p. 214); O'Connell (2016, p. 300); Pereira et al. (2021, p. 271).

[69] Ajibade et al. (2017); Arigbabu (2013); Côté-Roy and Moser (2019); Gabay (2018); Hoelscher et al. (2023).

[70] Bina et al. (2020); Cairns et al. (2022); van Noorloos and Kloosterboer (2018); Watson (2013); cf. Branch and Mampilly (2015).

[71] Myers (2018, p. 466).

[72] Myers (2018, p. 471).

Cities and urban environments do represent, for many, the possibility of diverse, multicultural communities living together in plural heterogeneity and creating a shared history and future.[73] Modern, tolerant, inclusive cities are often asserted or imagined to be safer for women, disabled, neuro-divergent, and LGBTQ+ communities than rural places. These assumptions buttress what Rahul Rao (drawing on Jasbir K. Puar) terms homonationalist and homocapitalist imaginaries, in which the acceptance of human rights norms and liberal tolerance of diverse others has become emblematic of the contemporary standard of 'civilization' and signifies 'a rosy future redolent with growth and productivity'.[74] Many liberals, critical theorists and progressive radicals still want to believe that building a bright, shiny and clean 'New Jerusalem' is possible.[75] A significant element of the political appeal of the *Black Panther* films was the Wakandan vision of a hyper-modern, urban African city-state, a place of shining towers, surging monorails, gleaming laboratories, and colourful markets, living in apparent harmony with the surrounding fields, forests and waters. An essay in *Time* magazine on how 'Black Panther could help us save the planet' praised its portrayal of a modern city without cars, sidewalks, or suburbs.[76]

Black Panther reflects how utopian or smart city narratives often feature claims for ecological sustainability and resilience as emblems of their modernity.[77] The Nigerian Long Term Vision proclaims that 'Nigerian cities will . . . play a major role in the country's climate change mitigation efforts', and envisions 'urban forms that are resilient to the effects of climate change', including 'agripolis urban farming systems'.[78] Prominent real-world examples include Eko Atlantic City on Victoria Island, Lagos, a luxury high-rise development built on land reclaimed from the ocean and protected by an 8.5km sea wall. It has been endorsed by the Clinton Foundation for its efforts to mitigate the effects of climate change in dense urban areas, but criticized by others for 'botched environmental impact assessments', population displacement, and exclusionary resource-intensive housing that 'perpetuate unequal configurations of power and colonial ideals of modernity'.[79] Iheka explores how the work of Afrofuturist artist Olalekan Jeyifous has sought to reimagine Lagos' slums

[73] Coward (2009); Nyamnjoh (2017).
[74] Rao (2020, p. 11).
[75] Delany (2009, p. 25). See also Bina et al. (2020); Tozer and Klenk (2018).
[76] Pierre-Louis (2020).
[77] Iheka (2021, p. 220); Watson (2013, p. 217).
[78] Federal Ministry of Environment (2021, pp. 20, 25).
[79] Côté-Roy and Moser (2019, pp. 2398, 2404). See also Iheka (2021, p. 215) and Watson (2013, p. 222).

as 'towering megastructures outfitted with solar panels, connected with skywalks, and spruced with greenery' in a vision of a post-carbon urban world of 2050 that shows us an 'aspirational model of the city of the future'.[80] Other notable examples include Hope City, outside Accra, and Kigamboni City on the outskirts of Dar es Salaam.[81] Yet few of these developments appear to be meaningfully sustainable or ecologically just in practice, prioritizing automobility and expensive apartments. They often require forced removals of shack populations and slum dwellers. Utopian visions can be recast as more complex heterotopian imaginaries through climate fictions that draw attention to the conflicts, contradictions, and lived experiences of African cities.

A more complex portrayal of future African cities appears in Tochi Onyebuchi's novel *War Girls*, set in 2172 after climate change and nuclear war have rendered most of the world uninhabitable. The novel follows the sisters, Onyii and Ify, members of a rebel Biafran army fighting the Nigerian state over mineral-rich land. It opens in their high-tech camp as the girls awaken in a tent. The world outside is 'green and wet from rain' but the dawn chorus of birds is shattered by massive humanoid robot war machines.[82] Later in the novel, Ify is taken by the Nigerian army to Abuja: 'An entire city made out of glass', full of skyscrapers, mosques, high-speed railways, and digital technology. Ify

> dreams of building extraordinary structures that will beat back the waters that gobble up more and more shoreline with each passing month. And she will figure out ways to harness that energy and power entire cities with it. She will figure out how to terraform those parts of pasture in the North that the desert has conquered.. . . She will make Nigeria a beacon of light on the continent.[83]

Yet the city of light is the product of a war-machine, a hierarchical, tightly controlled society, and is filled with surveillance orbs..

Another fascinatingly complex portrayal of the tensions within an apparent eco-city utopia is envisioned by Terh Agbedeh in the short story, 'Mango Republic'. It is set in Lagos in 2060, in an era of extreme global warming where Lagos is an ecological haven, 'the envy of other cities all over the world . . . an eco-city model for Africa' and beyond.[84] The city is protected by a dome and filled with sustainable technologies, including solar shields and power, rapid

[80] Iheka (2021, pp. 217, 220). To view these images, see https://jeyifo.us/SMS (accessed 30 October 2024).
[81] Watson (2013, pp. 220–1).
[82] Onyebuchi (2019, p. 4).
[83] Onyebuchi (2019, pp. 124–5, 128).
[84] Agbedeh (2013, pp. 183, 197).

transit, rooftop farms, energy from biomass and sewage, water converted to energy and used for healing, and even zero gravity machines. It is ordered and well-governed, housing is allocated to all and only two children are permitted per family.[85]

The story concerns the life of Aromire, an orphan-turned Area Boy, turned scientist. After a tumultuous life things finally seem to be coming together as his wife is pregnant with their first child. This is a fragile and long-desired pregnancy as Aromire's wife Rakia has suffered seven miscarriages since they have been married. Aromire reflects that '[h]er womb seems to have been made with eggshells for it shatters too soon, most times as soon as she has weathered the storm of the first trimester, discarding the precious new life.'[86] This time things seem to be going well, but as the pregnancy goes to term and Rakia's waters break the incessant rains outside cause the flood barriers to fail and Lagos Island is washed away. The story opens with the line: 'It feels like the city's waters just broke.' Environmental disaster and civic ruination accompanies personal fulfilment. Aromire sees his wife and his city in parallel and highly gendered ways: he believes human greed and hubris has 'raped' and abused the city of Lagos, a city which is 'like a woman with a troubled pregnancy [who] has had a series of miscarriages in the past but that has not stopped the people who keep trooping to the city gates, built to curtail the inflow that had become a surge.' The city has always drawn 'suitors' such as Aromire's ancestors, who 'came for the sweetness and like everyone else, left her pregnant with rage.'[87]

The joy of a new birth is qualified by the fragility of the eco-city. The title of the story, 'Mango Republic', has connotations of sweet, sticky ripeness, but the stagnant waters in the aftermath of the flood are 'juicy, messy like an overripe mango waiting for that nudge to fall to the ground and rot.'[88] It is clear there is trouble in utopia. The floods were 'a long time coming but everyone failed to take note' and the deluge was 'reminiscent of the days of the Biblical Noah; only this time, no-one had built an ark.' Aromire and Rakia—both scientists—have been too wrapped up in their personal family drama to pay enough attention to the weather, just as the rest of the city has failed to prepare. 'Global warming and the havoc its neglect would bring upon the earth had become the new gospel but life went on. Denial was the new religion and everyone practiced it with hermit-like dexterity.'[89]

[85] Agbedeh (2013, p. 182).
[86] Agbedeh (2013, p. 193).
[87] Agbedeh (2013, pp. 181, 195–8).
[88] Agbedeh (2013, p. 198).
[89] Agbedeh (2013, pp. 181, 195).

One of the most interesting themes in this story concerns the relationship between the inside and the outside. The city of Lagos is sealed within a dome—a recurrent trope in speculative fiction—and its links to the hinterland are ambiguous: it is 'the envy of other cities', a refuge of social justice, yet migrants lay 'siege' to its gates.[90] Aromire is both an insider as an establishment scientist, and an outsider as an orphan descended from immigrants. Pregnancy itself is often emblematic of blurred boundaries and disruptions to conventional inside/outside distinctions. As Iris Marion Young observed, 'the birthing process entails the most extreme suspension of the bodily distinction between inner and outer.... In pregnancy I literally do not have a firm sense of where my body ends and the world begins.'[91] Thus, in 'Mango Republic' we can't help but reflect on the ambiguous parallels between Rakia's 'eggshell' womb and Lagos itself inside its fragile dome. Aromire concludes in the final line: 'Life can be ironic, Lagos Island is no more and the city is in mourning but I am celebrating the birth of my baby.'[92] On one reading this story might reassert the security of heteronormative reproductive futurism with the birth of a child in a secure, conventional family unit; yet the collapse of the city and society outside the family home suggests something rotten at the core of the utopian ideal.[93]

The relationship between inside and outside—and the implications of this for care-giving—is also a prominent theme in Alistair Mackay's novel *It Doesn't Have to Be This Way*.[94] Set in Cape Town over an extended period of time before, during and after 'The Change' when flooding turns Cape Town into an island, the novel follows four main characters—Luthando, Malcolm, Viwe, and Milo—and their intersecting storylines. In near-future Cape Town we join Luthando, Malcolm, and Viwe at an eco-activist tree-planting festival, where each of them feels like an outsider as a result of being queer and/or black. Intersecting arcs of inclusion and exclusion in this novel—race, class, gender, sexuality, mental health, and political activism—show how cities are more or less open and accommodating to different communities, and are often actively hostile places for large swathes of their inhabitants. By the end of the book these power relations have been violently exacerbated by climate breakdown and a sealed compound called The Citadel has been built on Signal Hill and Lion's Head. Those inside live comfortable, if boring lives of work and shopping, increasingly filtered through virtual worlds and biotech implants.

[90] Agbedeh (2013, pp. 183, 194). See also Bina et al. (2020, p. 9).
[91] Young (1984, p. 49). See alsoKristeva (1986) and Ní Mhurchú (2016).
[92] Agbedeh (2013, p. 199).
[93] Edelman (2004).
[94] Mackay (2022).

Those outside are barely able to survive the heat, the lack of safe food and water, and the violence of poverty and desperation.

The story is a fairly heavy-handed warning about what contemporary climate politics could lead to. As Luthando and Malcolm's virtual reality climate activist films proclaim: 'it doesn't have to be this way' (#idhtbtw). There is some depth and nuance, however, in the intersecting character arcs and the ebb and flow of queer, caring relationships over the course of the story. In the end, the strength of the memory of these relationships leads Malcolm to leave The Citadel, reject his comfortable life and biotech implants and seek out Luthando and Milo. 'Luthando looks up at him when he hears Malcolm come into the room. The expression on his face makes everything worth it. "You've come home", he says.'[95] This reconciliation is described sparsely and it is clear there is no happy-ever-after ending. Malcolm is very sick and badly dehydrated, and Luthando and Noxolo, Milo's parents, go missing when trying to find food and water. Milo is left alone with the unconscious Malcolm, until he too leaves to try to find his parents. At the close of the novel it seems very likely that all the central characters either are or soon will be dead. But the temporary survival of Milo, the child, in the cave covered with his father's drawings—invoking South Africa's Blombos cave which contains some of the oldest known human cave markings—offers at least a chance of survival and hope.[96] Crucially, however, it is a hope that lies outside the sealed and gated compound of the city.

It Doesn't Have to Be This Way captures something that is true of many of the stories in Africanfuturist climate fiction. For the most part, these stories don't romanticize the cities of the future, rather they usually involve attempts to leave or flee these urban places in order to survive or in search of hope.[97] Even naïve Malcolm sees freedom in 'being able to upload human consciousness to the cloud' and escape from their 'meat cages'.[98] He later realizes that virtual reality worlds inside a sealed compound is just another sort of cage, and he seeks freedom by both leaving The Citadel and having his implants removed by a religious fanatic with a machete.

This theme of escape from the city is common to several other stories. In *Noor*, AO feels 'such a strong urge to flee where I was, who I was, why I was, when I was'. In her flight she is joined by a herder, DNA, also exiled from his home and they eventually find temporary safety in the desert, in a 'refugee city

[95] Mackay (2022).
[96] On the Blombos caves, see Wynter and McKittrick (2015, pp. 66–70).
[97] Cloete (2015, p. 369); Hoad (2016, p. 299); Mbembe (2019, p. 188); Simone and Pieterse (2017).
[98] Mackay (2022).

in the shadows.'[99] *Pumzi* ends with Asha leaving the compound to plant a tree in the wilderness, probably dying in the process but revealing that there is life beyond the sanitary, dream-suppressed and sealed city.[100] *Zoo City* ends with Zinzi leaving Johannesburg to 'get lost for a while' and track down Benoît's family, which she thinks might be 'the best thing I've done with my miserable life.'[101] *Mara and Dann* have to keep moving and Dann sometimes thinks 'that all the people alive are on their feet walking somewhere.'[102] Those who stay in cities like Chelops are only those who are unable or unwilling to acknowledge the reality of the coming drought, and when Mara and Dann finally reach their goal, The Centre, it is a ruin sinking into the marshes. Mara wonders about this: '[r]uined cities. Cities of all kinds. What was it, why was it, this law that beautiful cities had to fall into ruins?' Eventually she concludes that '[c]ities were as temporary as dreams.'[103] Revolutionary transformation comes to Lagos from outside in *Lagoon*, and the President has to travel out of the city into the water to meet the alien elders in order to bring the future to life.[104] This is a future that will spread beyond Lagos, notwithstanding the special role of the city and its human and nonhuman inhabitants in ushering in change. Even the *Black Panther* films advocate for a cosmopolitan politics of global engagement and solidarity, rather than remaining within the safe and secure walls of the utopian Wakandan city-state.[105]

These stories thus help to contest and complexify idealistic and naïve visions of smart African eco-cities that will usher in modernity and sustainability. These heterotopian stories insist on keeping power, violence, and inequality in sight, and show how architecture and urban design might facilitate new forms of inclusion and exclusion. The city as a sealed enclave is a utopian dream, and one that is always premised on the exclusion or violent marginalization of others. Heterotopian climate imaginaries envision futures where the relation between the city and the hinterland is in flux, and where lines of escape and flight are necessary and inevitable.[106]

Imaginations of future urban environments are so important for climate politics because they force us to consider where and how people will live. Who will we live with? How will domestic life be structured? How will food, water,

[99] Okorafor (2021, pp. 22, 130).
[100] Kahiu (2009).
[101] Beukes (2010, p. 358).
[102] Lessing (1999, p. 84).
[103] Lessing (1999, pp. 245, 361).
[104] Okorafor (2014).
[105] Coogler (2018, 2022).
[106] Mbembe (2004; 2019, p. 43). See also Davis (2006, p. 120); Jameson (2007, p. 15); Miéville (2016); Nielsen et al. (2021).

waste, transport, work, care, energy, etc. be organized? Of course, these questions arise in rural contexts as well but there is greater risk here that these settings can become wilderness imaginaries, stories of the struggles of frontier masculinity to survive in heroic pioneer landscapes. Imagined future cities—existing in mutually constitutive relations with hinterlands—raise inescapable questions about the future of family life, domesticity, and care.[107]

Homesteads of the Future

How and where will we care for others in climate-changed futures? In contrast to political imaginaries which almost completely neglect the domestic sphere, climate fiction has often offered compelling and intimate portrayals of the homes and care-giving of the future. The political economy of social reproduction—understood here in a narrow sense as 'having babies, raising children, providing care and immediate provisioning'—is a central albeit often neglected element of future transitions to sustainability.[108] The final part of this chapter explores how contemporary Africanfuturist climate fiction envisions future domestic spaces and care-full relations.

One of the strengths of fiction as a form—especially in contrast to policy texts and strategic visions—is its ability to portray individuals, families and caring relationships on a scale that is deep, believable, nuanced, and complex.[109] Many of the most well-known climate fiction novels are centrally concerned with caring relationships within family units. This ranges from the survivalist masculine fantasy of *The Road* through the dark satire of anti-feminist politics in *The Ice People* to the self-absorbed and neglectful parenting of *A Children's Bible* or *Solar*.[110] Kim Stanley Robinson's climate fiction is more concerned with world-building, political strategy, and techno-scientific imagination but, as Johns-Putra has shown, his portrayal of various types of future communal care-giving relationships illustrates the degree to which these questions must concern any ecopolitical imaginaries.[111]

A powerful vision of care, community, and alternative modes of parenting in the midst of climatic breakdown is offered by Octavia E. Butler in *Parable of*

[107] Bridge et al. (2013, p. 336); Davis (2010); Hajer and Versteeg (2019); Hoelscher et al. (2023); Johns-Putra (2019, p. 68); MacGregor (2006); Simone and Pieterse (2017).

[108] Luxton (2018, p. 37). See also Fakier and Cock (2018); MacGregor (2006); Paterson (2016); Swilling and Annecke (2012).

[109] Boswell (2014); Milkoreit (2016, p. 172); Nikoleris et al. (2017); Trexler (2015, p. 171).

[110] McCarthy (2006); Gee (1998); Millet (2020); McEwan (2010).

[111] Johns-Putra (2019, pp. 148–50).

the Sower.[112] Set in a near-future Southern California, devastated by social and environmental collapse, Lauren Oya Olamina seeks safety outside traditional family structures, eventually building the utopian enclave of the Earthseed community surrounded by an America wracked by corporate greed, inequality, and right-wing politics. Significantly, for Zamalin, Olamina's 'refusal to actualise a familial reunion between her and her long-lost daughter . . . refuse romantic gendered teleology' and turns the novel and its sequel into 'a radical statement about reorganising the nuclear family'.[113] Butler's interest in alternative family structures was taken even further in one of her most famous short-stories, 'Bloodchild', which she described as her 'pregnant man story'.[114] It explores symbiotic relations between species—humans and the insectoid Tlic—in the form of an unsettlingly strange love story where the human male, Gan, eventually chooses to host the eggs of his family's friend and nonhuman patron, T'Gatoi.[115]

Butler's novels and short stories have proved hugely influential for Afrofuturist and Africanfuturist climate fiction.[116] They encourage us to ask, what aspects of families and domestic life are worth preserving? And which aspects are damaging and dangerous? What family structures might help challenge patriarchy, homophobia, consumer capitalism, individualism, anthropocentrism, and racism? There is a huge diversity of responses within contemporary Africanfuturist climate fiction, as might be expected. Chinelo Onwualu's short story, 'Letters to My Mother', is set in a bucolic forested homestead in far-future Nigeria, where each settlement is 'uniquely suited to its environment' and where the only threat appears to be the traumatic memories of the past resurfacing.[117] The narrator, Obeche, describes how the home she shares with her life-partner, another woman, is 'nestled in the heart of an island in a river delta, never far from our forest. There were fruit trees and berry shrubs at every turn, and vines and creeping melons twined along walls and balconies. Even the moss underfoot was edible.' Obeche's compound is 'typical: five brightly painted homes arranged around a central courtyard. Each house was made of live bamboo saplings carefully pruned and bent toward each other until they formed a dome, leaving space for large windows. The walls were finished with hardpacked earth and the floors lined with smooth river stones so we were

[112] Butler (2019).
[113] Zamalin (2019, p. 133). See also Bollinger (2007, p. 343); Duncan (2005); Frazier (2016); Haraway (2016); Lewis (2019, p. 22).
[114] Butler (2005). See also Helford (1994, p. 263).
[115] Bollinger (2007, pp. 333–5); Jackson (2020).
[116] Womack (2013, p. 110); Zamalin (2019).
[117] Onwualu (2022, p. 67).

always cool in the heat.' Obeche finds a journal dating back as far as 'the climate wars', and its account of tensions between a mother and her daughter born from rape are so traumatic to read that it makes her sick. Obeche eventually manages, through meditation and communal dance and song, to turn the grief and trauma into 'a memory rich with life'. The story ends: 'I danced until I collapsed. And I was lucky, for when I fell, my mother's hands were there to catch me.'[118]

This account of a relatively safe and secure homestead is something of an outlier in Africanfuturist climate fiction. In many of the stories a central theme is the destabilization of the conventional narrative arc that associates family and the home with safety and security. This is not only a feature of literature; in conventional political imaginaries the image of the home—or homeland—is usually associated with safety and security, that which needs to be protected from dangerous outsiders.[119] In contrast, in many of these Africanfuturist climate stories families are a source of threat and instability. In another short story by Chinelo Onwualu, 'What the Dead Man Said', the family home is a place of trauma, not only of childhood rape by a trusted uncle but also the subsequent silence of the narrator's father.[120]

Another example is Okorafor's short story, 'Spider the Artist'. The central character, Eme, is living in the Niger Delta in a failing marriage. She believes the problem is that she has been unable to have children. Yet the wider context is crucial: the Delta oil pipelines are guarded by Anansi droids, dangerous artificial intelligence machines known as Zombies who secure the pipelines and respond to attacks, accidents, or trespassing with lethal violence. Eme's husband's violence, drinking, and depression are closely connected with the petro-state economy: '[m]y husband was a drunk, like too many of the members of the Niger Delta People's Movement. It was how they all controlled their anger and feelings of helplessness. The fish, shrimps and crayfish in the creeks were dying. Drinking the water shriveled women's wombs and eventually made men urinate blood.'[121]

Eme finds solace and temporary escape at the bottom of her garden, on the edge of town near the forest, 'just past the bushes, through the tall grass.'[122] Here she meets an Anansi droid. Remarkably, they seem to enjoy each other's company and the droid loves listening to Eme play guitar. Then, it makes its own musical instrument:

[118] Onwualu (2022, pp. 67, 79).
[119] Mbembe (2019); Weldes (2001).
[120] Onwualu (2019).
[121] Okorafor (2011). See also Kapstein (2016).
[122] Okorafor (2011).

> Suddenly, it rose up on its eight legs with a soft whir. It stretched and straightened its legs until it was standing a foot taller than normal. From under its body in the center, something whitish and metallic began to descend.. . . Right before my eyes, I watched the Zombie take this wire and do something with five of its legs while it supported itself on the other three. The legs scrambled around, working and weaving the shiny wire this way and that.

With the new instrument it plays music 'so complex and beautiful that I was reduced to tears of joy, awe, ecstasy'.[123] Playing music in the garden with the Zombie, Eme is finally happy.

> My Zombie had solved my marital problems. At least the worst of them. My husband could not beat me when there was beautiful music sending his senses to lush, sweet places. I began to hope. To hope for a baby. Hope that I would one day leave my house and wifely duties for a job as music teacher at the elementary school. Hope that my village would one day reap from the oil being reaped from it. And I dreamt about being embraced by deep blue liquid metal, webs of wire and music.

The story ends in disaster, however. The Zombies turn on their creators and start attacking people and hiding in the forests: 'men were torn limb from limb, their bloody remains scattered all over the swampy land'. In Eme's village there is an oil leak and when people gather the droids appear and set a spark to the spill, immolating the community including children and the school. Eme is sheltered by her droid whom she has named Udide, while the rest of the village perishes. 'I remember light, heat, the smell of burning hair and flesh and screams that melted to guttural gurgles. The noise was muffled. The stench was awful.' In the final lines of the story we learn that not only Eme but also her unborn child has been saved from the flames. It is a 'baby that my body allowed to happen because of Udide's lovely soothing music'.[124]

As well as a cyborg love story, this is a tale rooted in the crisis of family life. Eme tells us that '[m]y husband was beating the hell out of me during these months. I don't know why. He had not lost his job. I knew he was seeing other women. We were poor but we were not starving. Maybe it was because I couldn't bear him children. It is my fault I know, but what can I do?'[125] In the wider context of oil-dependent, violent, techno-authoritarian Nigeria, a normal family life seems impossible. The short story ends with a pointed warning

[123] Okorafor (2011).
[124] Okorafor (2011).
[125] Okorafor (2011).

to an international readership about the likelihood of this violence escalating: 'You should also pray that these Zombies don't build themselves some fins and travel across the ocean.'[126]

If 'Spider the Artist' ends with a new life amidst death, Totlo Tsamaase's short story, 'Virtual Snapshots', ends with a death as a result of a callous family betrayal. It is set in a futuristic Botswana where life is commoditized and largely digitized, under a dome 'created to protect us from the destructive environment we had orchestrated'.[127] The narrator spends most of her time inside DigiWorld, a virtual reality, because '[o]ur physical selves are latched to glass pistons by way of plastic tubes feeding medicine into our narrow veins'. Then she is summoned to visit her pregnant, sick mother 'outside'. She explains that '[o]utside DigiWorld, it is expensive to maintain our health, which is why when we partially disconnect we must pay fees to keep us breathing'. She makes the dangerous and difficult journey to see her mother but is met by an older sister who tells her to leave a present and then go home again, without seeing the rest of her family. Lacking funds or shelter, the return trip will be fatal. The narrator realizes this was the family's plan all along: the narrator is an adopted, surrogate child, not born of the mother. She is being abandoned to save the family's money and reputation. She begins the return journey but the story ends with the collapse of her body and a sanitized corporate email apologizing for the loss of vital services: 'Please hold for the next available customer advisor. All networks are busy. In case of emergency, please hold onto the nearest human for self-powering, explaining clearly your predicament to avoid violence and he/she shall be compensated within 7 days. Solar Power Corporations appreciates your patience. Goodbye.'[128] Her family is a threat, ultimately to her very existence.

She is an outcast, we are told, as a result of her artificial gestation 'shaped by glass and steel' in the Born Structure, a tower in the centre of the city. This was the case even before the final betrayal: '[t]hey kicked me out and never kept in touch'. She describes the family home as 'a place of warmth. A place I was thrown out from because I had reached the age of independency—because I was not from her womb. I had to fend for myself, a pariah unfit for their royal homestead... I am not born. I am a manufacture of the Born Structure.' This elicits sympathy from others in society but no-one saves her. 'My neighbour half-waves from her stoep until she realises what's happening. Her tear is the last grace I feel.'[129]

[126] Okorafor (2011).
[127] Tsamaase (2016).
[128] Tsamaase (2016).
[129] Tsamaase (2016).

This association between families and danger, between homesteads and violence, is recurrent across many of the stories. Mackay's novel, *It Doesn't Have to Be This Way*, is a story of domestic violence, rape, homophobia, and bullying.[130] Lessing's *Mara and Dann* novels begin with the children's exile from the family home, and the various potential replacement families they find on their journeys are always ultimately a threat to their survival. Mara resists the Kin's plans for her to get pregnant in Chelops and flees from a loving relationship with Meryx. She has a risky abortion when she realizes she is pregnant.[131] She associates birth control with freedom and there is an awful, shocking inevitability about the tragic start to the sequel.[132] In *Zoo City*, we learn that Zinzi went to prison for the murder of her brother, Thando, and is now estranged from her parents; Benoît's family is lost, perhaps dead; and the grisly denouement of the plot—unfolding in the nightmarish home of Odysseus Huron—involves S'bu killing his twin sister Songweza.[133] In *Lagoon* there are multiple examples of domestic violence and rape, and Adaora eventually decides that while she loves her children, her politics and career must come first because 'she had things to do that went beyond motherhood'.[134] Phoenix Okore kills her mother in her prison cell and her lover, Kofi, in the bedroom of his home.[135] In the sequel, *Who Fears Death*, Onyesonwu is the child of wartime rape and the plot centres on her quest to avenge her mother and kill her biological father.[136] In *Black Panther*, T'Chaka kills his younger brother N'Jobu; Killmonger challenges his cousin T'Challa to ritual combat, almost killing him (and killing Zuri); and eventually Killmonger himself is killed by T'Challa.[137] In 'More Sea than Tar' the homestead is fragile: 'when I look at all the holes in the house where the water comes through, I think that everything is falling apart.'[138] The narrator, Uti, 'can't fall asleep in this house of broken relationships'. It is his father's schemes that lead to the death of his brother, Joseph, and Uti's own near-fatality. Domestic violence and rape are prominent themes in the short stories 'Land of Light', 'Eclipse our Sins', 'What the Dead Man Said', and 'Letters to My Mother'.[139]

[130] Mackay (2022).
[131] Lessing (1999, pp. 225–6).
[132] Lessing (1999, p. 261; 2006).
[133] Beukes (2010).
[134] Okorafor (2014, p. 280).
[135] Okorafor (2015a, pp. 83, 207).
[136] Okorafor (2018).
[137] Coogler (2018).
[138] Ize-Iyamu (2019).
[139] Embleton (2017); Onwualu (2019, 2022); Tsamaase (2019).

As these examples show, escape, flight, and exile from the homestead is a central theme in much Africanfuturist climate fiction.[140] Okorafor's novel *Noor* concerns the attempt by the central characters AO and DNA to escape violent pursuers from the city. They eventually reach apparent safety in the Hour Glass city inside the Red Eye Storm, where they see that 'people lived right out in the open, not a house in sight'.[141] The sympathetic portrayal of nomadic lifestyles that pervades this story—as well as others by Okorafor—thus culminates in a heterotopian space where 'home' is a communal, open, sandy area.

One implication to draw from these stories is that traditional families and domestic spaces might require at least as much radical transformation as energy infrastructures in order to build safe, sustainable, just climate futures. Dominant narratives in climate politics often rely upon clichéd invocations of saving the planet for future generations, building clean and green homes, and models of activism that either ignore gender and social reproduction entirely, or invoke what Johns-Putra calls 'motherhood environmentalism'.[142] In contrast, these Africanfuturist stories suggest that heteronormative family structures and domestic politics—with their potential for violence and their implication in bourgeois capitalism and coloniality—might be social relations that have changed, will change, and should change in climate futures. Contrary to conservative fears that the breakdown of the traditional family unit will presage the collapse of society more generally, queer and feminist theorists encourage us to look for other, non-normative ways of living and caring together.[143] Seymour suggests that queer values entail 'caring not (just) about the individual, the family, or one's descendants, but about the Other species and persons to whom one has no immediate relations'.[144] She argues that these are the kinds of values essential for more ecologically sustainable futures. Cameron Butler elaborates that a 'queer ethic of care' is defined by its 'rejection of usefulness and (re)production as prerequisites to "earning" care', focussing instead on our mutually shared Otherness and the importance of building emotional, imaginative, and playful interpersonal relationships.[145] This can create spaces and places for 'bodies and landscapes to flourish in unique, weird, and unpredictable ways, rather than being managed into idealized conceptions of what will be useful or beautiful'.[146]

[140] Moonsamy (2020, p. 189).
[141] Okorafor (2021, p. 136).
[142] Johns-Putra (2019). See also Butler (2017); Death (2024b); Seymour (2013).
[143] Lewis (2019).
[144] Seymour (2018, p. 27).
[145] Butler (2017, p. 282).
[146] Butler (2017, p. 283).

These Africanfuturist climate fiction stories provide much more fertile soil for imagining diverse future care-full relations and dwelling places than official low-carbon policies and strategies, ranging from Okorafor's desert communes to Agbedeh's socialized nests and Onwualu's forest homesteads. These stories are testament to the resolute belief, the ineradicable hope, that even in the blasted futures of Mackay's sweltering Cape Town townships or Kahiu's nuclear wasteland, humans of the future will 'make kin in lines of inventive connection . . . rather than, or at least in addition to, godkin and genealogical and biogenetic family'.[147] They urge us to overthrow the family and build hospitable 'care commune[s] based on comradeship'.[148]

Conclusion

Much of the fun of reading speculative fiction comes from asking 'where in the world are we?' Often we read these stories for their ability to transport us to new worlds. For Rajaniemi, this making-sense of unfamiliar worlds can lead to 'a transcendent, goosebumpy moment that opens still greater vistas beyond the page, in the reader's imagination'.[149] This chapter has shown that exploring the strange places and unfamiliar landscapes, cities, and homes of Africanfuturist climate fiction can challenge some dominant discourses and assumptions, and can lead us to some important political implications.

First, these stories suggest that we can't rely on invocations of the familiar and the local to mobilize climate action. The landscapes of the future will be strange, different, heterotopian spaces that disrupt and disturb our maps of the world and will require new ways of being and relating to place. Environmental justice will require hybrid interventions and a willingness to engage with the uncanny and eerie, rather than a rear-guard defence of 'essentialist' localism.[150] As Bridge et al. observe, many people will experience the low carbon energy transition 'as the transformation of landscape', and this will be a 'simultaneously creative and destructive process that significantly changes how different places are related to each other, economically, politically and even culturally, and at a range of different scales'.[151] For Pearse, the expansion of renewable energy infrastructure over the coming decades is likely to

[147] Haraway (2016, pp. 1–2).
[148] Lewis (2019, p. 31).
[149] Rajaniemi (2017).
[150] Heise (2008).
[151] Bridge et al. (2013, pp. 335, 339).

mean that 'the re-production of space will be a key point of contention in eco-social struggles'.[152] These struggles will not only be over familiar and loved landscapes—national parks, bucolic rural farms, and garden cities—but will also take place within and over strange and unsettling places.

Moreover, by rendering the familiar strange and defamiliarizing places we thought we knew, African climate fiction can help us view our own surroundings and environments more critically. This can help to deconstruct dominant binaries, such as the tendency to see homelands as safe and faraway places as dangerous, or cities as either utopian salvation or dystopian nightmares. Africanfuturist climate fiction warns us to beware techno-modernist fantasies of smart, eco-cities. While urban theorists are fond of proclamations such as that the 'ecological genius of the city remains a vast, largely hidden power', this admirable utopianism can all too easily feed into the hyper-modernist fantasies of engineers and architects.[153] Such projects tend to work in the interests of the rich and powerful and produce new forms of exclusion and injustice. The stories examined in this chapter help contest the 'blinding power' of modernist 'smart-city and techno-city narratives', and lead us to try to imagine new forms of cities-and-hinterlands that are heterotopian, fluid, and porous, hospitable and in-flux.[154]

Third, the family is a social relation that has, should, and will change in climate-changed futures. The ethics and political economy of care will be absolutely central in the societies of the future, but rather than romanticizing traditional, heteronormative feminine care roles, or nuclear family bonds, we should politicize them.[155] This suggests, as Zamalin argues, that we need more 'irreverence toward the boundaries of communities and identities which can never be fully stabilised'.[156] We need to be clear that an environmental politics does not have to be grounded in a 'white, heterosexual, familial reproductivity', rather seeking to encourage the 'cognitive estrangement of reproduction and birth' in order to imagine new types of future families, based on comradeship, diversity, and lines of imaginative connection.[157] Many of the stories examined in this chapter highlight the disruptive impact of gestation and birth, which can work to unsettle and re-politicize the utopian imagination and contribute to the challenge of imagining more just and more sustainable future places.

[152] Pearse (2021, p. 958).

[153] Davis (2010, p. 43).

[154] Cairns et al. (2022, pp. 93–4) and Nyamnjoh (2017, p. 266), respectively. See also van Noorloos and Kloosterboer (2018).

[155] MacGregor (2006, p. 6); Peterson (2020).

[156] Zamalin (2019, pp. 140–1).

[157] Seymour (2013, p. 7) and Jackson (2020, p. 121), respectively.

5
Heterotopian Temporalities and Afrocentric Futures

The sands of time slip through our fingers however much we try to hold onto them. Temporality is a tricky thing. As one of Doris Lessing's characters comments: 'Time, time, ... Whenever you use the word time it's different'.[1] The image of time slipping away, dissolving the past, is vividly conveyed in *The Story of General Dann and Mara's Daughter, Griot and the Snow Dog*, when they discover a hidden Sand Library containing preserved books from the past. Dann becomes obsessed with them, reflecting that looking at these books is 'like looking into Time, and our eyes are not made for that'.[2] His men try to translate and understand them, but eventually they are forced to expose the books to the air in their rush to salvage something before the library sinks into the marshes. 'Dann, desperate, grief-stricken, grabbed up book after book and saw it disintegrate in his hands. Fragments of paper bronzed and darkened as the air took it. "And there goes the wisdom of a hundred civilisations", said Dann. "Look, there it goes. Going, going, gone."'[3]

This dreamlike scene is pervaded by a sense of loss and the failure to fully remember or preserve the wisdom of the past. In our ecopolitical imaginaries, climate change threatens the survival and preservation of the past, and there is never enough time. The usual way in which time is represented in dominant climate change narratives—both scientific/policy and fictional—is the pressing urgency of action, a sense that time is running out for humanity. As scientist Adrian Helmsley exclaims in the Hollywood apocalyptic disaster film, *2012*, 'I thought we'd have more time!'[4] In this chapter I argue that African-futurist climate fiction challenges dominant forms of temporality in climate change narratives. Time and temporality are complex concepts, more so than we sometimes assume. Everyone has a grasp of time passing, but as soon as we

[1] Lessing (2006, p. 265).
[2] Lessing (2006, p. 187).
[3] Lessing (2006, p. 250).
[4] Emmerich (2009). See also Luke (2015); Malm (2021); McKibben (2013); Rothe (2020); Wallace-Wells (2019); Whyte (2020).

African Climate Futures. Carl Death, Oxford University Press. © Carl Death (2025).
DOI: 10.1093/9780198960775.003.0005

start discussing or comparing time it emerges that we're often talking about different things or experiencing time very differently. Céline Granjou and Juan Francisco Salazar argue that various non-Western forms of temporality 'have too often been pushed to the fringes'.[5] In this chapter the forms of temporality represented in Africanfuturist climate fiction are brought to the centre of the story.

My own temporal perceptions have shifted while researching and writing this book. Caring for infant children, I became intensely aware of the time until the next feed, the next nap, how long they'd slept. How long I'd have to read or write before I had to pick them up from nursery. How many hours since the last dose of calpol. How long they'd spent watching TV or how many minutes more Peter Rabbit role-play I had to get through before I could let them watch TV. How many hours I had slept, compared to the last three days, seven days, two weeks. Time intensifies incredibly as you count the seconds between their night-time coughing fits or try to judge the best moment to put a child who has fallen asleep in your arms back to bed. Time flows differently at night than during the day, and a child's sense of time passing is very different to an adult's. My own temporal horizons have also expanded to encompass an altered relationship with my parents who are now grandparents, and my grandparents who are now both greater and lesser than I remember from my childhood. During the time I have been reading and writing for this book, all four of my grandparents have died, and both my cousin and a close school-friend were painfully eviscerated by cancer, robbing us of imagined shared futures together. Covid-19 tore families and society apart, to be painfully stitched back together by the few who had the time, patience, skill, and care to do so. And yes, I also worry about the climate-changed futures my children will face; the lure of heteronormative reproductive futurity is strong. During a heat-wave in West Africa in February 2024, Tarly, a carpenter in Abidjan, observed that '[w]hen it's this hot, mixed with humidity, time stands still'.[6] Time sputters and starts in unpredictable ways and our temporal horizons are conditioned by our situated perspectives.

Two important temporal imaginaries are those of linear and cyclical time. Linear time, we tend to believe, progresses incrementally, regularly, chronologically. Things happen one after another and are never exactly repeated: you can't step into the river of time in the same place twice. Linear, chronotic time underpins climate science in that past greenhouse gas emissions shape present

[5] Granjou and Salazar (2016, p. 242). See also Hutchings (2008, pp. 160–2); Iheka (2017, p. 17); Mitchell and Chaudhury (2020); Rao (2020).

[6] Quoted in Carrington and Borràs (2024).

climate conditions, and actions by powerful 'modern' characters in the present shape predicted future trajectories.[7] Grand narratives of modernization and progress often appear to invoke linear time, as the march of History proceeds apace, sweeping away that which went before.[8] Yet even dominant linear conceptions of time are more complex than is often imagined. Modernity arrives in leaps and bounds, there are tipping points, set-backs, cataclysms, and transformations. Some times are chronotic, signifying repetitive continual progress; some times are kairotic, signifying a time of action, upheaval and revolution.[9]

An alternative view of time is cyclical. Most of us also have an appreciation of what this means. Seasons come and go and come round again. Children grow up, mature, perhaps have their own children and pass on. Generational time involves loops and repetitions, at the levels of genetic inheritance as well as social organization, even as we know that nothing really stays the same. Climate change appears to pose a threat to the predictable repetition of the seasons, the rains, the migration of birds and beasts, the safety of our descendants. We are so often told we need to take action to address climate change now to save our children and grandchildren.[10] Linear and cyclical notions of time are both, therefore, part of dominant climate temporalities.

In contrast, the novels, short stories, and films examined in this chapter suggest more complicated and convoluted relationships between past, present, and future. Of course, there are moments in which linear or cyclical formulations appear; neither of these are as simple or as easily dismissed as they sometimes seem in critical discussions.[11] But it is a core attribute of Africanfuturist climate fiction that the passing of time is rendered more complex, more unpredictable and more reflexive than in dominant climate temporalities. In this respect Africanfuturist climate fiction is part of a much broader tradition of science fiction in which dominant temporalities are subverted.[12] As many have pointed out, it is a common misunderstanding that science fiction is about imagining or predicting the future. Rather, in the classic formulation of Frederic Jameson, the essential critical role of science fiction is to 'defamiliarize and restructure our experience of our own *present*', to transform 'our own present into the determinate past of something yet to come'.[13] Science fiction is about the future but not 'as a specific chronological register'.[14] It is

[7] Jasanoff (2010, p. 241).
[8] Ghosh (2016, p. 123); Táíwò (2022, pp. 198–9).
[9] Hom (2020); Hutchings (2008); Lundborg (2016).
[10] Butler (2017); Johns-Putra (2019).
[11] Hom (2020).
[12] Edkins (2013, p. 289); Mitchell and Chaudhury (2020); Smith (2012).
[13] Jameson (2007, pp. 286–8).
[14] Freedman (2000, p. 55). See also Thaler (2022, p. 10).

meaningless to ask how temporally far away an imagined the future has to be before it becomes science fiction. Rather, the future is 'a locus of radical alterity to the mundane status quo', so that our own present 'is thus estranged and historicized as the concrete past of potential future'.[15] As a genre of fiction explicitly concerned with reflecting on and subverting notions of time and temporality in order to defamiliarize the present, science fiction has proved a useful resource and inspiration for theorists exploring temporality and global politics. Postcolonial theorists, feminists, and queer theorists have all made important contributions to destabilizing the hegemonic imaginaries of white, patriarchal, straight time.[16]

In this chapter I explore four ways in which African climate fiction challenges dominant climate temporalities. First, many stories imagine Afrocentric global futures, in which the African continent is at the centre of global politics. This turns conventional assumptions about Africa and progress on their head. In the dominant modernization narrative, famously encapsulated as 'first in Europe then elsewhere', Africa is at best required to emulate technological advancements pioneered in Europe, North America, or China.[17] At worst, Africa is abandoned to the eco-apocalypse and Africans are trapped living dystopian lives that are nasty, brutish, and short.[18] In contrast, these Africanfuturist stories imagine Africa at the forefront of global civilization in a climate-changed future.

Second, the stories discussed in this chapter tend to represent time as heterotopic, multiple, and enfolded, rather than either predictably cyclical, or linear and *straight*.[19] Unlike dominant scientific narratives—whether the cause-effect of greenhouse gas emissions or the contested 'new era' of the Anthropocene—the future in Africanfuturism does not straightforwardly unfold as a consequence of past and present events.[20] Instead, in these stories, time loops and spirals, there are multiple co-existing 'nows' and futures that reshape the past. These are stories about trauma, in which the horrors of the past are repeated and re-experienced; stories about dreams which stand aside from the conventional flow of time; stories about eschatological and apocalyptical 'end of times' where the destruction of the world is played out again and again.[21]

[15] Freedman (2000, p. 55).

[16] Some of the most influential theorists of temporality and global politics, for my research, include Chakrabarty (2012); Edelman (2004); Evans (2017); Ghosh (2016); Grosz (2004); Hom (2020); Hutchings (2008); Kafer (2013); Mitchell and Chaudhury (2020); Rao (2020); Rothe (2020); Whyte (2018).

[17] Chakrabarty (2000, p. 7).

[18] Gabay (2018, pp. 1–2); Mbembe (2016, 2017); Rao (2020, p. 1).

[19] Muñoz (2009, p. 83).

[20] Franz (2022, pp. 777–8); Gergan et al. (2020); Yusoff (2018).

[21] Gibson (2020); Rothe (2020); Whyte (2018).

Third, these stories insist on the importance of attending to the traces of the past in the present and future, often focusing on the troubling necessity of remembering. In contrast to dominant narratives of environmental modernization in which the challenge of climate change is another legacy of past mistakes to be swept away through innovation and advanced technologies, these stories draw our attention to the ruins, ghosts, remnants, and echoes of previous technologies, societies, and economies. It is not surprising that many science-fictional texts feature strange museums, given their attention to defamiliarizing our own pasts and presents.[22] The detritus of the past, both good and bad, plays an active role in these stories, reflecting a postcolonial world order in which colonial legacies structure global futures, and indigenous histories and cultures survive and mutate in hybrid ways. In contrast to how dominant narratives assume that past 'mistakes' such as colonialism, homophobia, and racism will be swept away alongside polluting energy systems, these stories explore how the old gods, ancient artefacts and ongoing structures of coloniality and Whiteness continue to structure the present and the future. Paradoxically, they are also part of a utopian tradition within science fiction that asserts that 'radical difference is possible and that a break is necessary', rejecting the 'universal ideological conviction that no alternative is possible, that there is no alternative to the system'.[23] This utopian tradition has particular force and appeal within Afrofuturist and Black theory, as well as within queer traditions where theorists like José Esteban Muñoz search for concrete utopias within the quotidian struggles of the here and now.[24]

Fourth, the stories resolutely confront the 'problem of Time', in terms of the omnipresence of death and decay.[25] Death is necessary for life; necropolitics and biopolitics are constitutively intertwined.[26] If dominant narratives imply an emancipated future that suggests the promise of immortality, Africanfuturist stories remind us that the mortality of our bodies—and our civilizations—is a fundamental premise of socio-climatic futures. Doris Lessing's characters in the *Mara and Dann* novels exemplify this preoccupation with time and decay. Near the end of the sequel Dann reflects that '[h]owever you measure it, time will swallow the Centre, whether by water or fire, or both'.[27] The integration of such an apparently bleak understanding of time as death within an overall

[22] Jameson (2007, p. 99).
[23] Jameson (2007, pp. 231–2). See also Thaler (2022, p. 75).
[24] Fanon (2008, p. 5); Womack (2013); Zamalin (2019). See also Muñoz (2009).
[25] Hom (2020, p. 11).
[26] Mbembe (2019).
[27] Lessing (2006, p. 265).

politics of utopian possibility is one of the principal political contributions of Africanfuturist climate fiction.

This chapter discusses a range of novels, short stories and films from across the genre of Africanfuturist climate fiction. The novels of Nnedi Okorafor are particularly important for challenging dominant temporal narratives; *Lagoon* and *Who Fears Death* are both examined closely. The *Black Panther* films are also significant in imagining an Afrocentric global future, and Wanuri Kahiu's short film *Pumzi* narrates heterotemporal futures in interesting ways. This chapter also engages closely with the *Mara and Dann* novels by Doris Lessing. Her relationship with colonialism and postcolonialism—as well as with communism and feminism—are contested, controversial, and have fascinating implications for rethinking temporality.[28] Engaging with her *Mara and Dann* novels thus invites a discussion of white authors and wider structures of Whiteness in relation to Africanfuturism, climate change, and science fiction.

Afrocentric Global Futures

In much political discourse, as well as scientific perspectives and popular literature on climate futures, 'Africa'—usually homogenized and treated as a singular entity—often functions as a gloomy prognosis of the global future.[29] Africa's imagined present, rendered in terms of cyclical and worsening drought, poverty, violence, forced migration, famine, and political/environmental disaster, is held up as a cautionary tale of what lies ahead for more prosperous regions if they fail to mitigate and adapt to climate change. In a dismal reversal of more optimistic discourses of progress and Enlightenment, Africa represents a future dystopian and neo-Malthusian nightmare where human survival is imperilled by an increasingly hostile and inhospitable environment.[30]

In some ecopolitical imaginaries it is explicitly or implicitly assumed that by the point the rest of world is confronted with African-style climatic catastrophe, Africa itself will have been rendered inhospitable. In Kim Stanley Robinson's influential climate fiction novels Africa is often already dead or dying. In *Red Mars*, for example, there is one brief mention of the fact that by the mid twenty-first century, 'Africa was dying, and the north bickered over who they should help first.'[31] Similarly, Naomi Oreskes and Erik Conway's

[28] Bertelsen (1986); Primorac (2016); Sperlinger (2017); Tayeb (2009); Watkins (2010).
[29] McKibben (2019); Pereira et al. (2021).
[30] Gabay (2018, p. 122); Howell (2019); Mbembe (2016); Nyamnjoh (2017); Zamalin (2019, p. 12).
[31] Robinson (1992, p. 532). See also Schneider-Mayerson (2019).

'future history' tells the story of how a 'second Dark Age had fallen on Western civilisation' and climate change accelerated from the 2020s, told from the perspective of a Chinese superpower in 2393.[32] They report that 'social order began to break down in the 2050s, governments were overthrown, particularly in Africa'. Sea levels rise by 8m and 20% of the global population is displaced; 'human populations of Australia and Africa, of course, were wiped out'.[33] In other versions of this dominant narrative Africa is largely irrelevant because it will remain a region of interminable cyclical disasters and a struggle for bare survival.[34] In both cases, Africa is peripheral to or absent from global socio-climatic futures.

In contrast, Africanfuturist climate fiction narrates Afrocentric global futures. The most well-known example of this comes in *Black Panther* and its sequel where the long-hidden central African kingdom of Wakanda emerges from its self-imposed isolation to offer a divided world a prospect of a new golden age. This vision is technologically driven in the form of Wakanda's (extractive and non-renewable) vibranium-based civilization, but Wakanda also offers a broader socio-political vision of political tolerance, justice and restitution, forgiveness, and living in balance with nature and ancestral tradition.[35] As such the film has been heralded as an 'iconic cultural moment' for the African American community, proclaimed to be 'the coming of age of Afro-futurism' and bearing the weight of the 'brilliant possibilities of the past, present and future for African peoples on the continent and in the Diaspora'.[36]

There are many other examples of Afrocentric global futures in the Africanfuturist genre. Indeed, in some respects this is a defining feature of Africanfuturism: that the future is imagined in African-centric terms.[37] Tade Thompson's very successful *Rosewater* trilogy, for example, has the battle for the future between humanity and alien invaders being staged in Nigeria, while Britain is defeated and America is cut-off and incommunicable.[38] Neither *Black Panther* nor the *Rosewater* novels are primarily climate fiction, but they take place in a future shaped by climate change, and the relationship between humanity and nature is a central theme of their narratives.

The best examples of Afrocentric global futures in climate fiction come in the work of Nnedi Okorafor. First, in her petrofiction novel *Lagoon*, alien

[32] Oreskes and Conway (2014, p. ix).
[33] Oreskes and Conway (2014, pp. 25, 30, 33). See discussion in Mitchell and Chaudhury (2020).
[34] Wallace-Wells (2019). See also Iheka (2021, p. 10).
[35] Coogler (2018, 2022). See also Pereira et al. (2021); Pierre-Louis (2020).
[36] Zeleza (2018).
[37] Eshun (2003); Hugo (2017); Okorafor (2019); Womack (2013).
[38] Thompson (2016, 2019a, 2019b).

visitors visit Lagos to 'bring you together and refuel your future.'[39] The utopian conclusion sees the ailing Nigerian president revived to health and vigour by the aliens, and he is inspired by the offer of their technological assistance to assemble an elite team of advisors to lead Nigeria and the region into a new era of sustainability and progress. The rest of the world watches, jealously. This is nothing less than 'the potential birthing of a new world order' of post-oil civilization, a socio-climatic imaginary in which the 'solution' to carboniferous capitalism comes through welcoming and being hospitable to the radical Other.[40]

Nnedi Okorafor's novels *Who Fears Death* and its prequel, *The Book of Phoenix*, are mythic tales in which Africa is the site of recovery and rebirth after the ecological apocalypse. *The Book of Phoenix* tells the story of the collapse of the West, of how 'the wildest, darkest, rabid scientific sorcery' of twenty-first century corporate American imperialism led to climate change, injustice, slavery, violence, and genocide.[41] It opens in a flooded Manhattan of palm trees and mangoes. Vines grow on the skyscrapers to keep the air breathable and scientific research is focused on 'methods of dealing with climate change and buoy technology for floating towns and cities.' Behind 'the good intentions and amazing science, however, was abomination. Weapons, the quest for immortality, how far we could go. . . . The foundation of all the towers was always always always corrupt, driven by a lusty greed.'[42] The terrifying heroine, Phoenix, is born in one of these towers in a scientific laboratory. She escapes and is given a seed by a tree called the Backbone, which she takes and plants in Wulugu, Northern Ghana. After her capture and forced return to America she immolates modern civilization in a climatic apocalypse. Her story is recorded and taken to Africa by her friend, Saeed, who walks across the dried-up oceans. The sequel, *Who Fears Death*, is entirely set in a far-future Sudan. The redemption glimpsed at the end suggests that Onyesonwu holds the key to a future human civilization that could live in balance with nature. In that sense these novels offer, as Lisa Dowdall suggests, a distinctively Afrocentric utopian vision.[43]

Another powerful example of an Afrocentric global future, this time in the form of a contrapuntal climate narrative, comes from Doris Lessing. Her epic novels *Mara and Dann* and *The Story of General Dann and Mara's Daughter,*

[39] Okorafor (2014, p. 113). See also Jue (2017, p. 172).
[40] Hugo (2017, p. 48).
[41] Okorafor (2015a, p. 170).
[42] Okorafor (2015a, p. 98).
[43] Dowdall (2013).

Griot and the Snow Dog are survivalist dystopia in which southern African drought and a new European ice age have trapped central and north African societies of the far future, here called Ifrik, into endemic war and forced migration. In contrast to Oreskes and Conway's vision of the future, Africa is where life and civilization have survived while European cities lie buried under ice. We never really find out when the story is set; it could be 20,000 years after the ice came, or longer, but no one knows.[44] Importantly, Ifrik has experience of surviving climatic change. One character, Felix, tells Mara and Dann that '[n]othing stays the same.... Now it is a time of dryness. But the drought will end. And we will be prepared. All the history of Ifrik has been that—swings of climate.'[45] For Dennis Walder, Lessing's *Mara and Dann* expresses 'nostalgia for the future, a future in which Africa becomes the centre of the world, a place where new life may spring after the catastrophes anticipated in so much of her writing'.[46]

These Afrocentric global futures challenge dominant narratives that routinely ignore or render African futures invisible or explicitly extinguish African life. They insist that Africans have a central place in climate-changed futures. Yet, while they challenge aspects of dominant narratives, they also invoke another well-known storyline in which Africa is a source of renewal for Western civilization.[47] In this discourse, Africa's wildness, its status as the 'cradle of civilization' from which homo sapiens emerged, its art and aesthetics, and the survival in Africa of nomadic and indigenous human cultures, are all fused together to produce a symbolic Africa that becomes the source of human—and particularly White/Eurocentric—rejuvenation.[48] This storyline underpins the trajectory of Maggie Gee's novel *The Ice People*, which parallels and prefigures Lessing's *Mara and Dann* novels as it tells the story of the coming of the ice to Europe and the fleeing of the central characters south from England through Europe towards Africa where they believe sanctuary awaits them.[49]

Another example of this romanticized view of African salvation is the Roland Emmerich eco-apocalypse film, *2012*. Following a worldwide disaster caused by unprecedented solar flares heating up the Earth's core and causing earthquakes, volcanoes, tectonic shifts, and mega-tsunamis, the central characters—a white, middle-class all-American family, led by their novelist father—escape in a plane which, in heavy-handed symbolism, is named

[44] Lessing (1999, p. 199).
[45] Lessing (1999, p. 374).
[46] Walder (2008, p. 104).
[47] Mbembe (2017, pp. 26, 41–3).
[48] Gabay (2018, pp. 33, 65, 223); Mbembe (2016).
[49] Gee (1998). See also Johns-Putra (2019, p. 79).

'Western Spirit'.[50] They use their connections as well as unbelievable luck to join the few wealthy elite who have purchased sanctuary on board massive floating arks. Vivid apocalyptic spectacles such as this film reach huge audiences and shape eco-disaster imaginaries in powerful and commercially successful ways.[51] Eventually, the principal characters arrive on the coast of Southern Africa, where they see the Drakensberg mountains, the new 'roof of the world'. Africa, we are told, probably never even flooded as '[t]he entire African continent has risen'. Captain Michaels, in charge of one of the arks, remarks '[t]hat's why they call it the Cape of Good Hope'.[52] The film ends with a retreating camera shot of the Earth in space, centred on the African continent. In the post-apocalyptic future in which the rest of the world has been destroyed, Africa—once the Cradle of Civilization—offers hope for humanity's future. The (white American) family will find a new home out here. Strikingly, there is no mention at this point, or any other point during the film, of what has happened to Africans during the planetary apocalypse. There are no African leaders, diplomats, cities, or elites involved in the ark project. The erasure of Africans in this film is partially offset by the prominence of African-American characters, notably the President, his daughter and the heroic scientist Helmsley, all represented positively in contrast to the many selfish, cruel, greedy, and insane white characters.[53] As Helmsley and the president's daughter predictably 'get together' at the end of the film, we are invited to see Africa as the setting for an African-American return to the promised land and the site for the rejuvenation of human civilization. But Africa itself is a blank slate and actual African people and societies are erased from the future. This is yet another example of how, as Clive Gabay observes, 'Africa stands as a receptacle, long held to be devoid of its own history and culture both absolutely, and in comparison, to other non-Western territories, into which the genius of Western civilisation can be poured and sustained.'[54]

These examples show how the political potential of Afrocentric global futures is ambiguous. For Gee and Emmerich it is African landscape and climate that renders it central to a new world order. This is also the case for Lessing's *Mara and Dann* novels, although this is further complicated by an unresolved and ongoing conflict within the novels over the degree to which African societies are actually able to adapt and respond to climatic changes.

[50] Emmerich (2009).
[51] Joo (2020).
[52] Emmerich (2009).
[53] Joo (2020, p. 75).
[54] Gabay (2018, p. 35).

Black Panther proudly proclaims African culture and values at the heart of a new global civilization, although this is one driven by extraction of a fantastical non-renewable resource, vibranium.[55] It is in Okorafor's novels that the vision of an Afrocentric global future is most radical, revolving around the claim that African cultures and African agency have the inherent potential—albeit contested, problematic, and resisted—to craft better futures for human-nature relations.

Heterotopic Time

A storyline in which Africa offers a post-apocalyptic haven might be counter-intuitive for Western modernization discourses, but it does not always challenge dominant narratives as we have seen above. This is especially the case in storylines that proceed in linear fashion from stability through crisis to some kind of new stability or sanctuary. However, many of the stories told in Africanfuturist climate fiction challenge these linear narratives, instead imagining temporalities that loop and spiral, formations of 'broken time' with multiple co-existing 'nows' and futures that reshape the past.[56] In these stories, the passing of time is not smooth and even but has 'ruptures, nicks, cuts ... instances of dislocation', untimely moments that bring with them 'the difference that portends the future'.[57] For Cajetan Iheka, Afrofuturism is characterized by 'temporal entanglements' that reject linearity in favour of layering different temporal conjunctures.[58]

An excellent example of this is the short film *Pumzi*, set in East Africa 'after the Water War'.[59] The future and past reinsert themselves in the film's present in various ways, most notably in the dreams and visions of the central character, Asha. The insular society of the Maitu compound is apparently threatened by dreams and seeks to suppress them through compulsory drugs. Asha refuses to take hers, allowing visions of the past and future to shape her decisions. For Iheka, 'Asha's dream is crucial to the film's ecological theme, useful for mapping a better life, and for offering a glimpse of futurity. The suppression of the dream is therefore tantamount to maintaining the status quo to the detriment of an alternative future.'[60] Asha also works in a museum which, for Mich Nyawalo,

[55] Iheka (2021, p. 2).
[56] Ashcroft (2017, p. 93); Womack (2013, p. 154).
[57] Grosz (2004, pp. 5, 11).
[58] Iheka (2021, p. 27).
[59] Kahiu (2009).
[60] Iheka (2021, p. 33). See also Samatar (2017, p. 184).

drawing on Foucault, functions as a 'heterotopia of time because of its impetus to collect everything, to stop time'.[61] It is within the museum that Asha finds a living seed, experiences dreams of the outside and the future, and ultimately seeks to 'forge a new beginning'.[62] The dream-like episodes of the film, the shifting legacies of the past, and the open ambiguity of the ending all mean that the temporality of the story is more heterotopic than linear.

Temporally confusing dreams and a complex relationship between past, present, and future also unsettles the characters in Lessing's *Mara and Dann* novels. The struggle to remember and the risks and rewards of forgetting are a prominent theme, emerging from the very first line: 'The scene that the child, then the girl, then the young woman tried so hard to remember was clear enough in its beginnings.'[63] For the rest of the book Mara tries to remember her past and what she has been told; it is a 'central mystery' of the novel for Tom Sperlinger.[64] In contrast, her brother Dann resists: 'in his mind was a door shut fast against memories'.[65] Initially, it is very hard for the reader to locate the novel in time or place; the setting is clearly rural and seems more prehistoric than futuristic. Lessing explores the way in which genetic legacies from the past continue to shape the present and future, and the author's note introducing the novel reflects on how the Neanderthals have bequeathed to us their diversity, their ability to live in any climate, and their endurance.[66] The first page of the novel, immediately after this author's note, contrasts the Mahondi and the Rock People, where the latter are clearly intended to resemble Neanderthals. This is confirmed as the novel reaches its conclusion in The Centre's museum, where Mara sees a tableau of Neanderthal life.[67] Early brief references to strangely durable brown tunics and then a 'cart-bird' suggest we are more adrift in space and time than we might have initially thought, but it is not until Daima—a Memory, a character whose role is to remember and retell history—tells Mara about the past civilizations and their collapse, that we realize the novel is set in the far future.[68] Diverse human cultures, fashions, and ethnicities disappear and reappear at odd moments during the story: when they reach the town of Kanaz in the north Mara is surprised to find it 'was populated by a people with lean, flat bodies and sharp features. Mara had seen

[61] Nyawalo (2016, p. 219).
[62] Nyawalo (2016, p. 219). See also Omelsky (2014); Rico (2017).
[63] Lessing (1999, p. 1).
[64] Lessing (1999, pp. 5, 7, 45, 369–70); Sperlinger (2017, p. 302).
[65] Lessing (1999, p. 63).
[66] Lessing (1999, p. viii).
[67] Lessing (1999, p. 378).
[68] Lessing (1999, p. 37).

them before, on the walls of the ruins near the Rock Village. And here they were, just as if thousands of years and many migrations had not come and gone.'[69]

There is no rest or peace for any of the characters, who are driven to continually move onwards by their past. Mara and Dann are pursued by Kulik, an old adversary from their childhood, who 'haunts them both' throughout the two novels.[70] Trauma and madness stalk the siblings and threaten to overwhelm them or turn them against their closest friends and each other. Despite leaving home aged seven, a few weeks later Mara realizes she can't remember her mother's face. Mara gets sick near the end of the novel and suffers from fever dreams: '[s]ometimes when she woke she did not know whether she was in this wretched marsh-side inn, in a room where you could see the sky, or back in the Rock Village'.[71] Dann provides the clearest examples of this temporal dislocation and the reader is introduced to the origins of his trauma very early in the story. They meet two similar looking men, one good and one bad, but they become fused in Dann's subconscious as singular: 'the bad one'. 'How was it possible? Mara thought. Could one night, one terrible night in a child's life, mark him for ever? So that he would never get free of it? Even though he couldn't—or wouldn't—remember it?'[72] When Dann gets ill his fever means he 'seemed to be reliving several different times in his past, and Mara could recognize from what he said, or moaned, or shouted, that he went back and back again to the Tower, and his suffering at those moments was terrible'.[73] Many of the symptoms of trauma complicate linear notions of time, including nightmares, flash-backs, hallucinations, depression, and paranoia. This has led Jenny Edkins to talk about 'trauma time ... a time that does not run smoothly, and a time where there is no past, present and future, but only one time, the time of the now'.[74]

Lessing's work is intimately concerned with dreams and dream time. She has spoken about how large sections of the plot of both *Mara and Dann* and its sequel came to her in dreams, as if her unconscious was writing the story. Much of *Mara and Dann* was inspired by her own memories of caring for her younger brother when they were children, a brother whom she didn't get on with at all when they grew up. In this respect the writing and the dream memories were quite painful, because 'you can invite memories you don't

[69] Lessing (1999, p. 337).
[70] Lessing (1999, p. 401; 2006, pp. 76, 226).
[71] Lessing (1999, pp. 52, 359).
[72] Lessing (1999, pp. 137, 194; 2006, p. 274).
[73] Lessing (1999, p. 296).
[74] Edkins (2013, p. 286). See also Iheka (2021, pp. 109–10); Punter (2016); Kaplan (2016); Samatar (2017, p. 180).

particularly want'.[75] Her characters are often dreaming: in fevers, flashbacks, memories of the past, and visions of the future. Dreams are also used as a metaphor for the instability of the present in a context of war and climate change: '[c]ities were as temporary as dreams. Like people.'[76] Lessing's interest in Sufi beliefs and Eastern philosophies are briefly alluded to through the character of Ali: both he and Dann believe they have met before, maybe as brothers in a previous life.[77] Much like the trope of the museum, dreams are a recurrent feature of much science fiction where they often serve to interrogate and destabilize 'normal' temporality. For Darko Suvin, science fiction itself could be described as a 'realistic dream' or 'formalised daydreams'.[78]

The shifts in temporal style between *Mara and Dann* and the sequel, *The Story of General Dann and Mara's Daughter, Griot and the Snow Dog*, also illustrate the heterotemporal dimensions of Lessing's novels. A sequel is usually a continuation of the story and characters, but in several important ways Lessing's second Ifrik novel changes tack.[79] Whereas the central plot of *Mara and Dann* is linear—the siblings journey north from their birthplace, the village of Rustam in southern Ifrik, to meet their apparent destiny in The Centre—the plot of the sequel is cyclical and often meandering or spiralling. Dann, now the sole central character, tries to walk around the Middle Sea but is distracted by a pleasant life on the islands for several years before he returns to The Centre. He is doomed to repeat his mistakes, re-tread his steps, and revisit his trauma. If *Mara and Dann* is driven by threat, urgency, and dynamism, the sequel is characterized by inertia, lethargy, frustration, madness, and an incessant return to the past. In *Mara and Dann* the heroes surge past the other characters and the wider population, moving faster and further than the rest; in the sequel Dann stands adrift amidst the flows of refugees from conflict to the east and the drought to the south. Even the snow dogs from the north move with more purpose and energy than he does. In *Mara and Dann* chapter breaks punctuate the narrative, usually signalling a new phase of the journey or the separation of brother and sister; in the sequel there are no chapters and Mara's absence hangs over the whole story. Lessing's writing is famous for innovations in form and structure; the contrast between these two books demonstrates her willingness to challenge the conventions of the sequel, the adventure-quest genre, and the epic saga.[80]

[75] Lessing (2017).
[76] Lessing (1999, p. 361).
[77] Lessing (2006, pp. 271; also 42, 171). See also Sperlinger (2017, p. 308); Watkins (2010, pp. 26–7).
[78] Suvin (2016, pp. xviii, 9–10).
[79] Watkins (2006, pp. 255–6).
[80] Pathak (2013); Sperlinger (2017); Watkins (2006; 2010).

Nnedi Okorafor's stories also offer some complex representations of heterotemporality. The novel *Who Fears Death* revolves around the foretelling of Onyesonwu's death, a prophecy that makes the future a moment in her past, meaning she lives haunted by the 'ghosts of the future'.[81] Onyesonwu's rewriting of the Great Book at the climax of *Who Fears Death* leads to waves of change rippling out and altering the past, present, and future. 'What I can tell you is that the book and all that it touched and then all that touched what it touched and so on, everything in that small sandstone hut began to shift. Not to the wilderness, that wouldn't have scared me. Someplace else. I dare say a pocket in time, a slit in time and space.' This 'rewriting' means that, as Onyesonwu says, '[t]he curse of the Okeke has been lifted. It never existed, sha.'[82] The power to rewrite history, it will turn out, is the power to reshape the present and the future. The final passages of the text suggest a new past and present is coming into being, one where men and women, and Nuru, Okeke, and Ewu live convivially alongside each other.[83] But the final lines are an admonition to a baying mob waiting for Onyesonwu's blood: 'Let them wait. They will be waiting for a long long time.'[84] By making them wait, this terrible heroine exercises her power over time, and in so doing she also confounds the prophecy of her death that has hung over the novel.

This feat of escapology is extended further by another act of rewriting. Okorafor continues beyond the epilogue, providing an alternative vision of Onyesonwu's fate in a final chapter entitled 'Chapter 1: Rewritten'.[85] In this version Onyesonwu escapes and survives. The book thus presents two parallel endings: 'one that is foreshadowed throughout the novel, Onyesonwu being buried to the neck in sand and stoned to death, and one of utopian hope, escape, and living on'.[86] These are not alternative endings between which the reader is invited to somehow choose, as is implicit in some climate policy scenarios, but rather these endings are parallel and indivisible. Onyesonwu and Okorafor rewrite the past, present, and future, twist the flow of time into new shapes. In an era in which the imminent, looming threat of climate change seems to exert an inexorable temporal pull, an ability to manipulate past futures and future presents is, to say the least, *timely*.

Time moves in unusual ways in the prequel, *The Book of Phoenix*, itself written after *Who Fears Death*. The prequel tells the story of a genetically

[81] Okorafor (2018, p. 187). See also Kotecki (2020).
[82] Okorafor (2018, p. 378).
[83] Okorafor (2018, p. 386). See also Burnett (2015); Kotecki (2020); O'Connell (2016).
[84] Okorafor (2018, p. 386).
[85] Okorafor (2018, p. 384).
[86] Hoydis (2017, p. 180). See also Dowdall (2013) and Pahl (2018).

engineered 'SpeciMen' named Phoenix, who is two years old but artificially accelerated to a biological age in her thirties. She decides to escape after Saeed, another prisoner in the tower, apparently commits suicide after watching a reconstructed film about the Holocaust. The towers are engaged in genetic research, based on DNA taken from Lucy, a ten-year-old from Ethiopia and a traceable descendant of 'mitochondrial Eve' who is engineered to be immortal. Another prisoner in the Towers is HeLa, also immortal, a genetic descendant of the real historical character Henrietta Lacks.[87] Her blood has been purchased by seven rich men to give them everlasting life. In these novels by Okorafor the past can be rewritten and the future shapes the present. The power to manipulate the flow of time is what defines her terrible heroines, as well as characters like the mysterious sorcerer Baba Sola who appears in several Okorafor novels and is described as one who 'moves backwards in time'.[88]

In *Noor* the central characters, AO and DNA, escape into the desert and find refuge in a hidden city known as the Hour Glass. When they enter they are instructed that '[e]very hour at 1:11, all clocks reset to 12:11, all satellite locations and communication, all passwords reset, as well. We call that The Rotation of the Glass. We live within the chaos.'[89] The radical changes that take place in the novel *Lagoon* also involve a rewriting of the past to produce a new future. The emergence of the alien Ayodele from the sea is described in a way that explicitly invokes Mami Wata, a West African water deity, thus fusing the old and the new. The aliens in *Lagoon* explicitly echo Dogon tales of the Nommos, an amphibious people apparently from the Sirius star system.[90] At the end of the novel we learn that the narrator is Udide Okwanka, a legendary spider from Igbo mythology, weaver of tales, creator of being. This fusion of the old and the new can also be seen in the description of 'the Bone Collector', a spirit that haunts the Lagos–Benin Expressway and devours travellers. Tradition (spirits) and modernity (the road) are thus bound together, and with the coming of the aliens—the novum—they lead the way to a new utopian future.[91] For Esthie Hugo, 'the value of Okorafor's African science fiction narrative hinges on its project of reversing the temporality of Western modernity', which she argues is essential for formulating better future-imaginings.[92]

These examples show how a complicated, non-linear relationship between pasts, presents, and futures is a key feature of much Africanfuturist literature.

[87] Okorafor (2015a, pp. 97, 186). See also Weheliye (2014, p. 79); Womack (2013, pp. 35–6).
[88] Okorafor (2021, p. 94).
[89] Okorafor (2021, p. 132).
[90] Hugo (2017, pp. 49–50); Jue (2017, pp. 175, 184); Womack (2013, pp. 85–6).
[91] Hugo (2017, pp. 52–3); O'Connell (2016); Suvin (2016).
[92] Hugo (2017, p. 56).

Traumatic memories from the past—and ghosts who speak to the living—also resurface in short stories by Chinelo Onwualu, Stephen Embleton, and Jessica Wilson, as discussed further in Chapters 4 and 7.[93] In Mackay's novel, *It Doesn't Have to Be This Way*, the characters Luthando and Malcolm debate how best to deal with memories of pain and trauma. In the final third of the book, after climate catastrophes have divided society into those living AI-enhanced lives of pleasure and diversion within sealed compounds like The Citadel and those barely surviving outside, Malcolm has continued his career as a software developer. He has created a mental health app that allows people to 'identify and neutralise their traumas ... delete the memories that haunt them'. In letters exchanged with Luthando, living outside the dome, Malcolm is warned against creating emotion-filtering software. Luthando asks him: 'don't you think it's important that people feel what they feel?' Malcolm is eventually convinced that 'he shouldn't be cauterising their traumas', especially when he realizes a private security company is willing to fund the app for counter-terrorism purposes. He then sabotages his own coding and flees the Citadel. Outside, 'the buildings are dark. They have been returned to candles, to fire, to sewerage thrown out of windows. The clocks turn back. History rewinds. Time ebbs and flows like the tides.'[94]

The past is not fixed in these stories, and it actively intervenes in and is reshaped by the present and the future. This suggests that policy interventions and democratic deliberations about climate futures would be enriched by opening up room for multiple temporalities and diverse, heterotemporal imaginaries of past, present, and future. In particular, imagined climate futures need to pay much more attention to the role that the detritus of the past plays in the present.

Detritus of the Past

The past is usually *that which needs to be overcome* in modernization narratives. Environmental modernization is no different: historic structures of resource use and GHG emissions need to be swept aside and replaced with modern, efficient, and sustainable technologies and forms of socio-economic organization. Invoking the charged metaphor of a locomotive—a powerful symbol of linear forward progress—the key challenge for environmental

[93] Onwualu (2019, 2022); Embleton (2017); Wilson (2019).
[94] Mackay (2022).

modernization is the 'decoupling' of economic growth from resource use and GHG emissions.[95]

The past tends to play a very different role in Africanfuturist stories. These stories draw our attention to the ruins, ghosts, remnants, traces, and echoes of previous technologies and cultures, showing how the detritus of the past plays an active role in the present and the future. This 'persistence of the past in the present' constitutes a foundation for radical politics, since it 'produces the resources for multiple futures, for open pathways, for indeterminable consequences, as well as for those regularities and norms that currently prevail'.[96] A vivid, if perplexing, example of this is the iconic Afrofuturist film *Crumbs*, which unfolds in a fantastically far-future Ethiopia where baffling quests take place amidst the plastic detritus of long-gone consumer civilizations. Any attempt to create new, better futures must deal with the active presence of traces of the past.[97]

Wanuri Kahiu's short film *Pumzi* also vividly portrays the surface of the Earth strewn with detritus. In contrast to the antiseptic cleanliness of the futuristic compound interior, when Asha leaves we see the rubbish ejected from the compound via a pressurized tube.[98] The landscape is uninhabitable, scarred by nuclear and toxic waste, presumably legacies from the water war. Inside the compound, objects from the past are confined to the museum where Asha is a curator. She takes an old metal compass when she escapes to search for life outside. Traces of the past play more symbolic and intangible roles as well. The film invokes Kikuyu myths and makes 'allusions to earlier films by African women', all suggesting 'that the future cannot be contemplated without recourse to the past'.[99]

Relics of previous civilizations litter Okorafor's novel, *Who Fears Death*. There are caves full of decrepit computers from long-gone 'mad scientists', the 'old devices of a doomed people'.[100] Some of these technologies remain in use: water capture stations, laser knives which can cut cleanly and cauterize wounds, and solar powered technology including maps. There are also indigenous traditions and cultural practices, such as a circumcision rite that Onyesonwu's mother describes as 'barbaric'.[101] Inherited legacies are also part

[95] Baldwin (2012); Hindess (2007); Hom and Steele (2016); Rothe (2020, p. 151); Thaler (2022, p. 147).
[96] Grosz (2004, pp. 252–3). See also Hutchings (2008, pp. 67–8); Iheka (2021); Womack (2013).
[97] Llansó (2015).
[98] Kahiu (2009). See also Omelsky (2014, p. 40).
[99] Iheka (2021, pp. 29, 35). See also Sides (2019).
[100] Okorafor (2018, pp. 329, 334–7).
[101] Okorafor (2018, p. 45). See also Womack (2013, p. 113).

of the detritus of the past: Onyesonwu realizes she has the 'potential and taste for amazing cruelty', a 'gift from my father, from his corrupted genetics'.[102]

Lessing's *Mara and Dann* novels are quite similar to the world of *Who Fears Death* in this respect, full of the 'detritus and debris' left by previous civilizations, including technological, cultural, and genetic legacies.[103] Mara is fascinated with ruins outside the Rock Village where there are 'layers of habitations, peoples, time'.[104] Ifrik is full of these ruins and some of their more durable advanced technologies remain: indestructible but ugly brown tunics, metals that can't be scratched or bent. There are also technologies slowly falling apart from lack of parts and know-how: sky skimmers, automatic lights, solar panels, and railways. Early in their adventure they find a cave with the remnants of extinct animals: the adults with them are intrigued and urge Mara to remember where the cave is and come back to study it. But the cave is only mentioned once again when Mara recalls she has never been back.[105] She never will. Mara and Dann are repeatedly saved on their journeys by ancient technology they don't understand: indestructible water cans, the durable tunics, gold coins, a beautiful axe, sky skimmers, a solar powered barge, and a railway carriage. Yet these remnants sadden and frustrate them more than inspire them; to Dann they are '[r]ubbish, ancient rubbish'.[106] The description of the solar panel on the barge is particularly striking: 'the square of metal, which was unknown now, something invented in the distant past and forgotten.... Just a square of metal, flat and thin, nothing to it, which you might kick aside or throw onto a rubbish heap; and yet it was something to make you feel awe or even terror, because this bit of nothing much, that looked as if it had come off a scrap heap, could make this boat move upstream day after day.'[107]

In contrast to Onyesonwu in *Who Fears Death*, who uses ancient technologies when necessary but envisions a slate wiped clean, a new world order freed from the violence and constraints of the past, Dann and Mara feel a nostalgic sense of loss over the knowledge of previous generations. The goal of their quest is The Centre, where 'all the records of the great past were'.[108] Yet when they arrive it is a gloomy, sad, crazed place, sinking into the marshes and enveloped in mist. It is inhabited by the deluded Felix and Felissa, whose dreams of reviving the glories of the past Mahondi empire are

[102] Okorafor (2018, p. 151).
[103] Lessing (2006, p. 21).
[104] Lessing (1999, p. 57).
[105] Lessing (1999, pp. 18, 73).
[106] Lessing (2006, p. 131).
[107] Lessing (1999, p. 231).
[108] Lessing (1999, p. 377).

clearly anachronistic, out-of-time. During the sequel, this atmosphere of baffled and frustrated nostalgia reaches the level of suicidal depression for Dann. He is 'ashamed, surrounded always by a past so much more clever and wonderful and rich than anything they had now. Always now you came up against *long ago ... long, long ago ... once there was ... once there were* people, cities and, above all, knowledge that had gone.' Dann is haunted by his past: he 'looked as if he were seeing ghosts, or hearing them'.[109]

This sense of nostalgia in Lessing's *Mara and Dann* novels has attracted considerable critical attention.[110] There are troubling political implications when a white, English author who grew up in relative privilege in colonial Rhodesia writes a story in which two heroic, royal children travel through a future Ifrik constantly expressing their dismay at the poverty, ignorance and crumbling civilizations around them, in contrast to the fragments of advanced technology and shards of classic literature inherited from crumbled Yerrup. Fiona Becket suggests that Lessing risks erasing, dismissing, or avoiding 'Africa' with her invention of Ifrik and nostalgia for its past.[111] Although Kaplan doesn't discuss Lessing directly, she regards nostalgia and melancholy as typically 'male fantasies' about the end of the world.[112] In contrast, Walder regards nostalgia as one of the most 'uncanny' and 'profound ways of linking past, present and future'.[113] It is worth unpacking the implications of the nostalgia of the *Mara and Dann* novels in a little more detail, as it goes to the core of Lessing's views on the relationship of past, present, and future, as well as raising important questions about racialization and Whiteness in her work.

Nostalgia for a vanished or receding past is a familiar emotion in much imperial and colonial literature.[114] A superficial reading of Lessing's *Mara and Dann* novels is that they are full of melancholy regret at the fall of the culture and technology of Yerrup and espouse the hope that Mara and Dann, bearers of the Mahondi royal blood, will revive this past civilization. This is more-or-less the reading provided by Becket, who critiques the 'irresistible draw of the "happily ever after"' in the novels, and the appeal to the nobility and supposed 'racial superiority' of Mara and Dann.[115] This critique is less convincing, however, when read as part of Lessing's wider body of work, and in

[109] Lessing (2006, pp. 24, 129; see also 130, 135, 171, 192, 200, 203).
[110] Becket (2009); Sperlinger (2017); Tayeb (2009); Walder (2008); Watkins (2006).
[111] Becket (2009, p. 139). See also Watkins (2010, p. 14).
[112] Kaplan (2016, p. 149).
[113] Walder (2008, p. 101).
[114] Ashcroft (2017).
[115] Becket (2009, pp. 135, 138, 140).

particular her often misleading, ambiguous or unreliable 'authorial voices'.[116] To put it simply, even if some characters and some sections of text espouse views that seem nostalgic for an older imperial age, this does not mean this is the only or primary political reading of the novels. A more insightful reading, I suggest, is that these novels are meditations on the role of the past and memory in the present and the future, and a reflection on what Lessing saw as the tendency for societies in flux to cling to myths of past glories. Lessing herself remarked that in the context of colonial Africa, '[t]he emotional impulse behind nearly all white writing is a nostalgia, a hunger, a reaching out for something lost; hard to define but instantly recognisable'.[117] The plot of *Mara and Dann* is described by Lessing as 'a reworking of a very old tale' and the books are driven by a curiosity about whether in the future all human civilization so far 'will amount to a sentence or a paragraph in a long history'.[118] There is nostalgia here, but there is also an insistence that life carries on and that new stories and new ways of telling stories are possible. Key characters in the novels play the role of oral storytellers, passing on the tales of the past to the future: Daima and other Memories in the Rock Village and Chelops, and the character Griot—whose name means a West African storyteller, royal advisor, singer, or oral historian—who sings the tales of Dann and his sister at the end of the sequel.

The apparent nostalgia of Lessing's *Mara and Dann* novels must also be read alongside the racial politics of the stories. Again, on a surface reading it is undeniable that there are unsettling and often unpleasant racialized language and ethnic stereotypes in the book.[119] Groups like the Rock People, Hadrons, Hennes, and others are described in ways that should make the reader very uncomfortable. Mara and Dann often encounter groups of identical people who they regard as sub-human, lacking individuality and reason: 'these creatures.... They are like animals ... those upturned dull faces, a multitudinous unity, a nightmare'. When they pass a regiment of Hennes soldiers, we are told that

> It was easy to imagine those robustly planted legs, stamp, stamp, stamp, as the legs of a single organism, perhaps one of those long, shiny, brown insects like worms, the length of a forearm, and their legs all moving together, like a rippling fringe. And the brown uniforms made a blur, like a long brown

[116] Watkins (2010, pp. 96, 118).
[117] Quoted in Walder (2008, p. 100).
[118] Lessing (1999, p. vii).
[119] Tayeb (2009).

> crawler which, if crushed, shows it is filled with a whitish ooze. Easy to believe too, that the Hennes were filled with it, and not with the healthy blood of real people.[120]

Yet a superficial reading of these passages as evidence of a racist author is to do Lessing a disservice. The repeated 'easy to imagine' and 'easy to believe' in the above quote suggests we should read this description carefully. In passages such as this, and others which cast doubt on the reliability of Mara and Dann's instinctive first reaction to foreigners or their ability to use facial appearance as a judge of character, I read Lessing as drawing attention to the human trait to mistrust and fear Others who look and behave differently.[121] The vividly insect-like descriptions above recall the use of the term 'cockroaches' to describe Tutsis in the build-up to the Rwandan genocide in April 1994, and it is *not hard to imagine* how this might have been in Lessing's mind as she worked on this novel, published in 1999.[122]

Skin colour is frequently mentioned in these novels, with Mara and Dann encountering peoples of various brown, yellow, and greenish tones. Notably, the skin colour of the central characters is very rarely mentioned, and it is only confirmed quite late in the novels that Mara and Dann are probably of light brown coloration. The first character described as black is Kira's husband in the River Towns near the Cong river. Here the brief description is also quite stereotypically racialized: 'He shone with pleasure in his handsome blackness.' On the very next page—hardly a coincidence—Mara meets her first ever white person. She reacts with shock to the skin of a hostile government spy from Goidel, an 'ugly white colour she had never seen before ... unhealthy white skin—repulsive.'[123] The racial politics of the novels become quite complex and white characters like Leta and Donna initially horrify Mara and Dann. Leta's skin is described as 'alien and disturbing.'[124] But they soon become trusted members of their close family. There are black characters, like Dalide, who are at least highly respected and insightful, if not exactly sympathetic allies. The sequel contains several characters from Kharab and beyond to the East, like Ali and Sabir, who are trusted, admired and liked. It is notable, however, that we never meet a sympathetic, admirable, or fully humanized Rock Person, Hadron, or Hennes character, which means that the dehumanized, insect-like

[120] Lessing (1999, pp. 206–7, 235).
[121] Lessing (1999, pp. 41, 116, 304, 314–15).
[122] Hintjens (1999).
[123] Lessing (1999, pp. 219–20).
[124] Lessing (1999, p. 320).

descriptions of them noted above are never properly challenged or interrogated. This does leave at least some parts of the book with a disconcerting racialized tone.[125]

Crucially, however, Lessing makes very clear that while she believes it is an inherent human trait to distrust others, the structures of racism and prejudice are contextual and historical and thereby can shift.[126] Mara's childhood disdain for the Rock People and belief in Mahondi superiority is challenged when they arrive in Chelops where the Hadrons—related to the Rock People—rule the city and the Mahondi are seen as 'an inferior race and had always been servants and slaves'. Mara comes to question her own views of Mahondi superiority, describing her people as 'inbred'.[127] Dann's unshakeable belief that those who look a certain way are 'bad men' is key to his psychotic madness that so often leads to disaster for him and his sister.

Significantly, there is also an explicit and repeated critique of Whiteness in these novels. This works on at least two levels. First, in terms of a simple reversal in which people of white skin colour are initially regarded with suspicion and disgust, as noted above; and second, in terms of an explicit critique of the grand narrative connecting white or European peoples to modernity and progress.[128] The Albs, the white descendants of refugees from Yerrup, are described as 'a small-minded people' who behave as if the rest of Ifrik doesn't exist. 'They have only one thought, that their grandchildren or great-grandchildren will return to Yerrup when the ice melts. And it is going at last, so it is said.'[129] Lessing's upbringing among colonial expatriates in Rhodesia, of whose constant complaints and moaning she was so critical, is alluded to here.[130] The Albs are figures of disgust but also a perverse fascination for most of the characters in the book and there is a strange extended scene in the sequel where Dann's soldiers request and queue up to touch Leta's blonde hair. The impulse is described as driven by curiosity and seems almost innocent but there is an edge of threat and desire. It is clear that Leta experiences the episode as humiliating and an unpleasant echo of her previous life as a prostitute.[131] It is hard not to read this as an inversion of contemporary tropes around 'black hair' as critiqued by Emma Dabiri in *Don't Touch My Hair*.[132]

[125] See also Howell (2019, p. 178); Watkins (2010, pp. 119, 124).
[126] Tayeb (2009, p. 24).
[127] Lessing (1999, pp. 138, 222).
[128] Sperlinger (2017, p. 308).
[129] Lessing (1999, p. 387).
[130] Walder (2008).
[131] Lessing (2006, pp. 164–8).
[132] Dabiri (2020).

Even more pointed is the wider condemnation of European history and culture in the novels. Mara is full of awe for the technologies of ancient Yerrup but it is also made very clear that these were broken, psychotic, suicidal cultures.

> But the end of the story in every building was war, and the ways of war became crueller and more terrible. ... it was thought that these ancient peoples would have ended their civilisations even if the ice had not arrived.... These were people who had no interest in the results of their actions. They killed out the animals. They poisoned the fish in the sea. They cut down the forests, so that country after country, once forested, became desert or arid. They spoiled everything they touched. There was probably something wrong with their brains. There are many historians who believe that these ancients richly deserved the punishment of the ice.[133]

We also learn that, when the Ice came the Albs 'ran from Yerrup like rats', destroying the societies they reached.[134]

This means that, despite some of the unpleasant and ambiguous language and tropes in the stories, I read the *Mara and Dann* novels as reflections on the persistence of racial tropes from the past into the present and future. Susan Watkins, drawing on Deleuze and Guattari, regards Lessing's work of this period 'as "minor" genres which ultimately critique racist narrative'.[135] Lessing claimed that her work was a commentary on what she saw as the general human tendency of 'a dominant group despising the rest'.[136] The detritus of the past, including racism, ethnic stereotypes, and imperial ruins continue to structure the present and the future. This can mean that reading Lessing's work is an uncomfortable experience, and others have articulated important critiques of her use of racialized tropes.[137] Whatever Lessing's intentions, my reading of the *Mara and Dann* novels is that they powerfully portray and critique structures of racial anxiety and white fragility concerning the loss of knowledge, the fall of European cities and peoples, and the 'passing of the civilisational baton' to African and Middle Eastern populations.[138] In this respect, Gabay argues that 'Africa has thus always served as a site where, inspired by White anxiety and fragility, people self-identifying with Whiteness (the belief in the universality of a mythologised European institutional genius) have

[133] Lessing (1999, p. 381).
[134] Lessing (2006, p. 191).
[135] Watkins (2010, p. 124).
[136] Bertelsen (1986, p. 138).
[137] See Bertelsen, quoted in Walder (2008, p. 108).
[138] Gabay (2018, pp. 2, 31) and Walder (2008, p. 109).

sought to remake better versions of the West'.[139] It does not need too much stretching to see the *Mara and Dann* novels in this light, as an apocalyptic tale emblematic of an anxious, pessimistic, white liberalism.

Yet Lessing also opens the way for the 'possible end of whiteness' and the construction of alternative worlds, notably through her unambiguous condemnation of European culture as warlike and ecologically destructive.[140] She famously rejected the title 'Dame', a member of the Order of the British Empire, remarking 'I spent a good part of my youth trying to undo the British Empire.'[141] What I take as crucial in Lessing's novels, as well as the other stories examined in this chapter, is how the detritus of the past plays an active role in these stories, reflecting a postcolonial world order in which colonial legacies structure global presents and futures, and indigenous histories and cultures survive and mutate in hybrid ways. Climate fiction, I suggest, is often deeply concerned with imagining, negotiating, and narrating white identities in a context of climate apartheid, and sometimes this takes the form of 'profound anxiety' about the prospects for the 'lifestyles, forms of governance, conditions and structures associated with whiteness'.[142] In many of these stories—from the fantasies of post-racial conviviality in *It Doesn't Have to Be This Way* or 'The Drought' to Okorafor's enigmatic sorcerer Baba Sola, 'a white man who lives outside of whiteness'—the relationship between the racialized identities and power structures of the past, and transformed climate-changed futures, is unstable and interwoven.[143]

Even *Black Panther*—which in many respects can be seen as an African inflection of a hyper-modernization narrative—seeks to fuse advanced technology with indigenous traditions, ritual combat, and the wisdom of tribal ancestors. Ingesting the sacred heart-shaped herb allows the Black Panther to access the Ancestral Plane, a dream-like realm where communication with dead ancestors is possible.[144] In contrast to how dominant narratives assume that past 'mistakes' such as colonialism, homophobia, and racism will be swept away alongside polluting energy systems, these Africanfuturist stories explore how ancient worlds and ongoing structures of coloniality and Whiteness continue to shape the present and the future. As Farhana Sultana argues, '[c]olonialism haunts the past, present, and future through climate'.[145]

[139] Gabay (2018, p. 33).
[140] Mitchell and Chaudhury (2020, p. 322).
[141] Watkins (2010, p. 164).
[142] Heron (2024) and Mitchell and Chaudhury (2020, p. 7).
[143] Mackay (2022); Okorafor (2021, p. 94); Wilson (2019).
[144] Coogler (2018).
[145] Sultana (2022, p. 10). See also Hoad (2016); Rao (2020); Tamale (2020).

This has implications for contemporary climate politics. We need to pay attention to the traces of the past in the present and future, and think about how to preserve, memorialize, and repair that which has been bequeathed to us. We need to ask: how will the museums and histories of the future narrate our present? The archivist Obeche, in the short story 'Letters to My Mother', describes how residents of this tranquil, utopian, futuristic homestead are 'encouraged to chronicle their thoughts and daily lives for eventual inclusion in the archive when they passed on'.[146] In a real-world example, researchers Raven and Stripple describe how they employed speculative methodologies in the creation of an immersive exhibition—the Museum of Carbon Ruins—to explore how Sweden in 2053 might memorialize the last few decades of the 'Fossil Era', including objects that had to be given up and what came to replace it. The exhibition normalizes 'future low-carbon lifestyles' while 'contemporary practices of consumption are framed as strange and unfamiliar'.[147] Novels, films, and short stories can—like museums and art galleries—work to defamiliarize and estrange us from conventional narratives about our past, present, and future, and this is an important way to challenge and repair the damage wrought by colonialism, racism, patriarchy, and other injustices.

Crucially, however, alongside this shared insistence on colonial continuities and the detritus of the past, Africanfuturist climate fiction also offers the promise of revolutionary change. This will be discussed in more depth in Chapter 7, 'Reparative Climate Politics and Reimagined Subjectivities', but at this point it is important to note that many of these stories imagine futures that are very different to the past. If dominant narratives of transitions are that they emerge from within contemporary capitalism, technology, and social organization, African climate fiction opens up the possibility that all this might be over-turned and new world orders can be imagined that rewrite past, present, and future. Africanfuturist climate stories invoke traditions of revolutionary utopianism, while also insisting that the past is not swept away in these stories but continues to shape the present.[148]

This might seem contradictory. I suggest that this close relationship between continuity and change, the detritus of the past and revolutionary utopia, is another manifestation of the heterotopian temporality that characterizes Africanfuturist climate fiction. This heterotopianism is influenced by—and can also contribute to—debates within Black Studies and queer theory. Alex Zamalin argues that black utopians, from Frederick Douglass through

[146] Onwualu (2022, p. 66).
[147] Raven and Stripple (2021, p. 230).
[148] Táíwò (2022, pp. 202–8).

Sojourner Truth and Martin Luther King, to Samuel R. Delany and Octavia E. Butler, have envisioned future societies as 'founded in postracism rather than postracialism', in which racial categories and identities would not disappear but would be freed 'from the fetters of white supremacy and racial violence'.[149] For José Esteban Muñoz, '[q]ueerness is essentially about the rejection of a here and now and an insistence on potentiality or concrete possibility for another world', or 'a backward glance that enacts a future vision'.[150] He explicitly links this argument to black utopianism when he suggests that '[t]he queerness of queer futurity, like the blackness of a black radical tradition, is a relational and collective modality of endurance and support'.[151] For theorists like these, the past plays an active role in the present and the future yet they also insist that another world is possible. As Elizabeth Grosz argues, '[o]nly if the present presents itself as fractured, cracked by the interventions of the past and the promise of the future, can the new be invented, welcomed, and affirmed'.[152] A fracture, a break, is an act of violence, however, and it is to the close inter-relationship between violence and possibility, death and life, that the final section of this chapter turns.

Necropolitics and Biopolitics

The fourth way in which these Africanfuturist stories challenge dominant modes of temporality concerns their confrontation with the 'Problem of Time' or the presence of death, decay, and atrophy.[153] If dominant narratives seem to imply an emancipated future that suggests the promise of immortality, Africanfuturist stories remind us that the mortality of our bodies—and our civilizations—is a fundamental premise of socio-climatic futures. The utopian thread noted above is predicated on death and sacrifice. The fact that death and life are also so central to these stories suggests that the inter-relationship between necropolitics and biopolitics is a central feature of Africanfuturist ecopolitical imaginaries.[154]

In 2003 Sir Martin Rees, scientist and UK Astronomer Royal, reflected that he thought 'the odds are no better than fifty-fifty that our present civilisation on Earth will survive to the end of the present century'.[155] It is claimed

[149] Zamalin (2019, p. 13).
[150] Muñoz (2009, pp. 1, 4).
[151] Muñoz (2009, p. 91).
[152] Grosz (2004, p. 261).
[153] Hom (2020, p. 11); Kermode (2000).
[154] Mbembe (2017, p. 49; 2019); Nyamnjoh (2017, p. 257).
[155] Quoted in Mitchell (2017, p. 3).

that we are living through our planet's 'sixth mass extinction event', this time driven by human action. Extinction and death, necropolitics and posthumanism, have weighty implications for the stories we tell about ourselves and the futures we imagine. Extinction is not simply obliterative: it is a creative process that generates diversity as well as being destructive.[156] Living well—and dying well—are paramount ethical challenges of existing alongside other species as I will discuss further in the next chapter.

Pumzi is one of the clearest representations of the mutual implication of death and life in Africanfuturist climate fiction. The final shots of the film show Asha lying in on the ground in the desert, giving her last water to the seed she has planted. It seems clear that she has sacrificed her life to a dream of the future.[157] Whilst some have critiqued this ending, it is portrayed as a voluntary act, a moment of political agency in the spirit of wider traditions of sacrifice and black mortality that see death as a political opportunity.[158]

The *Black Panther* films also foreground the necessity of death and sacrifice for life to go on.[159] The plot of the first film opens following the death of T'Chaka, meaning that T'Challa returns to Wakanda to claim his father's throne. It emerges that T'Chaka killed his younger brother N'Jobu for betraying Wakanda, leaving behind a son who would grow up to be Killmonger. Killmonger challenges T'Challa to ritual combat, almost killing him (and killing Zuri). Ultimately, Killmonger himself is killed by T'Challa. The sequel, *Black Panther: Wakanda Forever*, begins with T'Challa dying from a mystery illness. This plot-line emerged following the tragic death of actor Chadwick Boseman in 2020 aged 43 which gave rise to an emotional farewell to his film character. He is succeeded by his mother, Queen Ramonda, who herself drowns in the act of saving the gifted young African-American scientist, Riri Williams. As Iheka reflects, 'the struggle to control vibranium leaves behind bodies—of humans and of destroyed environments'.[160] Whilst some critics castigated the 'Eurocentric stereotypes' of tribal conflict, ritual combat, 'Tarzanian animalistic chants' and 'shields and spears and gyrations of old Hollywood films', this prominence of death and sacrifice is clearly influenced by traditions of Black political thought and necropolitics.[161]

[156] Mitchell (2016, pp. 23–4).
[157] Kahiu (2009). See analysis in Nyawalo (2016); Rico (2017); Sides (2019).
[158] See discussion in Iheka (2021, pp. 38–9) and in the next chapter.
[159] Coogler (2018, 2022).
[160] Iheka (2021, p. 223).
[161] One of the most pointed critiques is levelled by Zeleza (2018). Yet I see resonances between these films and elements of the political thought of Jackson (2020), Mbembe (2019), and Sexton (2016), especially on the foregrounding of violence, death, and sacrifice.

Okorafor is concerned with very similar themes of sacrifice, where life and death are intertwined. *The Book of Phoenix* begins with an epigram from Robert Hayden's poem *Middle Passage*: 'Voyage through death, to life upon these shores'.[162] As her name implies, Phoenix dies and is reborn multiple times during the novel. Slavery and its consequences are not buried in the past but shape the present and the future, as Phoenix links her imprisonment to those of 'other Africans on unwanted voyages' across the Atlantic.[163] As the novel moves to its conclusion Phoenix 'saw death all around me' and resolves to 'Kill everything. Everything should die. Let it all start from the beginning. In the right way.'[164] The end of the world for some is therefore 'the prerequisite for the possibility of imagining "living and breathing again" for others'.[165]

Who Fears Death is a novel in which death and violence are constant themes. Onyesonwu is a dangerous, troubled, counter-cultural heroine; an Ewu, or mixed-race child born of wartime rape, in a world torn between racialized conflict between the Okeke and Nuru, legitimized through a scriptural Great Book. The book begins with her adopted father's death and funeral, where Onyesonwu discovers her power and almost brings her father back to life. She is on a journey of self-discovery as well as a political quest to kill her biological father and end the genocide. In one incident, she angrily makes an entire town go back in time to witness the rape of her mother.[166]

The final confrontation has cataclysmic consequences. Onyesonwu's partner and lover Mwita is killed in the struggle and Onyesonwu responds by making herself pregnant. Through her shape-shifting powers she brings Mwita's sperm, still inside her, to her eggs. 'At the moment of conception a giant shock wave blasted from me, a shock wave like the one so long ago during my father's burial ceremony. It blew out the walls around me and the ceiling above me.'[167] The explosion almost kills her weakened father, Daib, but the consequences are awful. 'Every single male human in the central town of Durfa capable of impregnating a woman was dead. My actions had killed them.' Much like her father, Onyesonwu reflects, 'We both leave bodies in our wake. Fields of bodies.'[168] Furthermore, all the women in the area are simultaneously made pregnant. The awesome, terrible symbolism of childbearing as an act of war mirrors her own brutal conception. In the acknowledgements

[162] Okorafor (2015a, p. vi).
[163] Okorafor (2015a, p. 94).
[164] Okorafor (2015a, pp. 198–9).
[165] Yusoff (2018, pp. 12–13).
[166] Okorafor (2018, pp. 151, 155).
[167] Okorafor (2018, p. 368).
[168] Okorafor (2018, p. 370).

Okorafor references an article about 'weaponised rape' in Sudan in 2004 as creating 'the passageway through which Onyesonwu slipped into my world'.[169]

Lessing's *Mara and Dann* novels are also profoundly shaped by death. True to the generic features of adventure quests, minor characters are eaten alive, killed in explosions, murdered in hand-to-hand combat or die from exhaustion and starvation with stomach-churning regularity. As in all good coming-of-age quests the eponymous heroes are separated from their parents on the opening page, later to learn of their death. Dann is repeatedly brutalized physically and psychologically, incurring shocking scars. Lessing makes the reader endure a graphic account of the damage caused when he is repeatedly raped.[170] When Mara says she doesn't want to see another weapon ever again, Chabis asks, 'what sort of time do you think you are living in?'[171] Birth and death are tightly intertwined: the drought means birth rates have fallen across Ifrik and fertile women are being forced into breeding programmes. Yet Mara is aware that pregnancy means likely death in these circumstances. She narrowly escapes rape and chattel pregnancy several times, and aborts one child even though she loves the father. Neither Mara nor the reader is spared the sight of 'a tiny creature in its bloody mess'.[172] At the end of the first novel Mara tells Dann that they could never live happily and peacefully together as they would end up killing each other: '[w]e have been made violent'.[173]

Yet, the real kick in the guts comes at the start of the sequel with the sudden death of Mara, a genre-defying twist that leaves the reader distraught. This, 'at a stroke, attacks the reader's investment in the notion that the sequel particularly confirms the continuity of identity'.[174] The news is introduced early and obliquely as Dann leaves The Centre. 'He walked, thinking of Mara and the past. His mind was full of Mara, and of sorrow, though he had missed the news of her death. She had died giving birth. The messenger from the farm had come running to the Centre, but Dann had left.'[175] Dann does not find out for three years and the reader has to mourn Mara alone for 75 pages.

Death, dissolution, and decay are a wider feature of Lessing's world in these novels. Changing climates are bringing forth new species, but the dominant tone is of loss and things falling apart. Dann observes that '[o]nce it had taken

169 Okorafor (2018, p. 389). See also Death (2022a).
170 Lessing (1999, p. 192).
171 Lessing (1999, p. 400).
172 Lessing (1999, p. 225). See also Pathak (2013, p. 2229).
173 Lessing (1999, p. 403).
174 Watkins (2006, p. 256).
175 Lessing (2006, pp. 18–19).

generations for a city to sink down into the mud, but now, much less?'[176] The inevitability of death, conflict, and the dissolution of knowledge and civilization in the sands of time—and yet the injunction to go on living and telling stories—is what I think makes Lessing's novels such a powerful addition to Africanfuturist climate fiction.

The fusion of utopian visions of the future, the legacies of the past, and the suffering and mortality of the present seems to me to be one of the most distinctive, interesting, and imaginative elements of how Africanfuturist climate fiction challenges dominant forms of temporality in climate change politics. In these stories, death and life are co-constitutive and mutually reproducing. The archivist Obeche in Onwualu's 'Letters to My Mother' takes the pain from the letters she finds and 'shares it with her community and the forest'. She concludes that, '[l]ike the transformation from waste to compost [the] grief became a memory rich with life, a catalyst that would help me and my people grow stronger'.[177] One way to describe the way these storylines fuse death and life into utopian possibility might be in Donna Haraway's terms, in which we are all 'communities of compost'.[178] Viewed from this angle it is striking how little space there is for remembrance, grief, mortality, or compost in dominant ecopolitical imaginaries.

Conclusion

Nobel Peace Prize winner Wangari Maathai concludes her book on the 'challenges for Africa' with this summation: 'our work is reclamation—bringing back what is essential so we can move forward. Planting trees, speaking our languages, telling our stories, and not dismissing the lives of our ancestors'.[179] This concern with re-stitching pasts, presents, and futures together in new ways that challenge Western notions of temporality and progress suffuses Africanfuturist climate fiction.[180] In this chapter I have argued that such stories challenge dominant climate temporalities in four significant ways: by imagining Afrocentric global futures; through heterotopian temporal structures that interweave pasts, presents, and futures in multiple ways; by insisting on the active role of the detritus of the past in the presents and futures; and the predication of all this on the inevitability of death, decay, and suffering.

176 Lessing (2006, p. 12).
177 Onwualu (2022).
178 Haraway (2016).
179 Maathai (2009, p. 288).
180 See also Táíwò (2022, p. 202).

The political implications of reading these stories in this way can now be articulated more directly. First, in contrast to the mainstream climate fiction canon, these stories foreground Afrocentric futures and demand that Africans have a place in climate-changed futures. They interrogate Whiteness, trauma, and memory more vividly than authors like Atwood, Robinson, McCarthy, Kingsolver, and McEwan. This is a contribution to diversifying climate fiction and expanding its imaginative potential. There should be more space for remembrance, grief, mortality, and compost in dominant climate imaginaries.

Second, they suggest that political strategies and visions of the future need to allow space for plural and non-linear temporalities, with multiple co-existing 'nows' and futures that reshape the past. This could be an important way to democratize the process of future climate governance, as it also raises questions about how to preserve, memorialize, and repair that which has been bequeathed to us. This includes asking: how will the museums and histories of the future narrate our present? What is the role of memory, tradition, and even death, within a radical, utopian politics of climate justice and reparation?

It is also worth acknowledging that Africanfuturist climate fiction does not completely reject dominant climate temporalities. For example, a central concern in much climate fiction concerns the failure of societies to act in the face of environmental catastrophe, or what Haraway describes in the current moment as the 'Great Dithering'.[181] This theme is certainly present in Okorafor's *The Book of Phoenix* and *Lagoon*, as well as a core concern of Lessing's *Mara and Dann* novels. It is also present in several of the other examples of Africanfuturist climate fiction not discussed in this chapter, such as 'Mango Republic' where the narrator observes that the flooding of Lagos 'was a long time coming but everyone failed to take note'.[182] Such stories are a powerful way to insist that if we want to survive to tell new stories in liveable futures then we need to transform carboniferous capitalism, and soon. But radical, challenging forms of fiction must do more to unsettle assumptions than simply add their voices to those urging more action on climate mitigation and adaptation, however important this might be. This chapter has shown that one of the crucial functions of Africanfuturist climate fiction is to unsettle and defamiliarize our very notions of time and temporality.

Thinking through temporality also means thinking through racialization, racism, and Whiteness. Contrary to 'race-blind' narratives that blithely assume that racism is a legacy of the past to be left behind in the future, reading

[181] Haraway (2016, p. 144).
[182] Agbedeh (2013, p. 181).

Africanfuturist fiction raises questions about racialized futures. The stories examined in this chapter confront race and racism directly, in contrast to the sublimated anxiety about the future (and the ghosts of the past) that haunt much of dominant climate politics. I have argued in this chapter that these stories can help to defamiliarize and estrange us from conventional narratives about our past, present, and future, and this is an important way to challenge and repair the damage wrought by colonialism, racism, patriarchy, and other injustices. This is important in itself, I think, but it is also important for me, personally. I am a product of my own ancestors, including Anglo Saxons, Celts, and the giants of Ath, Methodist preachers and wartime evacuees, teachers and carpenters and nurses and civil servants, and that oil company. Through reading fiction I can enter—if only ever partially—worlds that are not my own, and try to hear different voices and learn different lessons to those I was brought up with.

There is nothing inevitable, however, about the potential of fiction to challenge dominant discourses. As Edkins observes, some stories create a picture of a world 'that is orderly and organized, a world where categories hold, where time is linear and moves steadily from moment to moment, and we know where we are'.[183] For the most part, the stories in this chapter do not do that. By challenging dominant representations of temporality they help make new ways of being and becoming more possible. As Elizabeth Grosz has argued, '[t]he more clearly we understand our temporal location as beings who straddle the past and the future without the security of a stable and abiding present, the more mobile our possibilities are, and the more transformation becomes conceivable'.[184] This has implications for human relations with other species, which is the focus of the next chapter, as well as for ecopolitical imaginaries more broadly.

[183] Edkins (2013, p. 286).
[184] Grosz (2004, p. 14).

6
Dangerous Ecologies and More-than-Human Conviviality

Our world is full of dragons, dinosaurs, unicorns, talking bears, and fairies. At least, my world is, and I suspect this is true for many families with young children. My children have an insatiable appetite for stories and their experiences in the more-than-human tales of Julia Donaldson, A. A. Milne, Michael Rosen, Enid Blyton, Roald Dahl, and—naturally—Walt Disney, mean that they are unperturbed by plots in which a princess and a young dragon decide to become flying doctors or an unlikely friendship forms between a lion cub, meerkat, and warthog. Our youngest child has a particular fascination for crocodiles. They crop up in a surprising number of children's stories, although perhaps, given the prominence of charismatic African megafauna in books and films for infants, this shouldn't be a surprise. Crocodiles in stories for small children tend to be sneaky, greedy, brave, and cheeky. They make a terrific 'snap'. In this light it is worth remembering that the 'disenchantment of the world' is something that recurs every day, as children are explicitly or implicitly encouraged to grow up, be sensible, and read serious, realist fiction.[1] By the time we are adults—and certainly while engaging in the serious business of climate politics and net zero strategies—it would be something of a shock to encounter a crocodile with agency. *Cave! Hic dragones*, indeed.[2]

Lauren Beukes' vividly more-than-human novel, *Zoo City*, features some intriguing and entertaining human–animal relationships, as we will see below, but none are as scene-stealing as a monstrous albino crocodile that looks '[l]ike a fucking dinosaur'.[3] In many stories, crocodiles represent terrifying prehistoric threats, including in the *Mara and Dann* novels, the short story 'More Sea than Tar', and even Disney's 1998 direct-to-video sequel, *The Lion King II: Simba's Pride*.[4] Australian philosopher Val Plumwood has recounted her own near-death experience with a crocodile, vividly recalling its 'beautiful

[1] Horkheimer and Adorno (2002, p. 1). See also Ghosh (2021, p. 204).
[2] Strange (1982).
[3] Beukes (2010, p. 333).
[4] Ize-Iyamu (2019); Lessing (1999, 2006); Rooney (1998).

African Climate Futures. Carl Death, Oxford University Press. © Carl Death (2025).
DOI: 10.1093/9780198960775.003.0006

golden-flecked eyes' and the impact of the attack on her work.[5] It changed the way she regarded herself and other species, encouraging her to 'honour our food and the more-than-food that all of us are'. It also modified her view of death. 'I still think there's no other world', she reflected, 'but I don't think the story ends with your death. The story passes on to the other life forms you nurture with your death, nurturing those who have nurtured you, in a chain of mutual life-giving.'[6] The struggle to live and die well, alongside nonhuman others such as crocodiles—or water dragons as Lessing calls them—is a central theme of many of these Africanfuturist climate stories, and it sets up a more challenging and imaginative representation of more-than-human politics than the typical stories of polar bears, trout, and butterflies that recur in mainstream climate fiction.[7]

This chapter argues that climate politics and net zero strategies are impoverished by the absence of more-than-human politics. It shows how Africanfuturist climate fiction can make an important contribution to reimagining co-dependencies between humans and animals, ethical bonds between people and forests, deserts and waters, and encounters with nonhuman spirits or posthuman alien civilizations. These teeming, fertile interrelationships—what Amitav Ghosh calls 'the urgent proximity of nonhuman presences'—are in stark contrast to the ways in which nonhumans are represented within most imagined futures.[8] Nonhumans are usually either absent or resources for human development, and an appeal to consider nonhuman agency in climate change strategies is likely to be regarded as inappropriate, frivolous, or absurd. Despite the resolute anthropocentrism of dominant climate discourses, there is increasing if belated interest in the more-than-human and posthuman in the discipline of International Politics.[9] In this chapter, nonhumans is used to refer to species or entities that are not human; more-than-human is shorthand for ontologies that emphasize the constitutive interdependence and lack of clear boundaries between humans and nonhumans; and posthuman is shorthand for ontologies that 'take seriously the reality that what is currently recognised as "humanity" has not always existed, and it may not exist in the future'.[10]

The discussion of African net zero visions in Chapter 2, 'Dominant Narratives in Long-term Strategies', showed that nonhumans are largely absent

[5] Plumwood (2004).
[6] Plumwood (2004).
[7] Lessing (1999, p. 26).
[8] Ghosh (2016, p. 5).
[9] Burke (2023); Cudworth and Hobden (2011, 2022); Fishel (2023); Fougner (2023); Mitchell (2016, 2017); Simoneti (2022); Strausz (2024); Tschakert (2022); Youatt (2014).
[10] Mitchell (2023, pp. 1–2).

from ecopolitical imaginaries. Anthropocentric visions of 'the world in 2050' focus on human economies, societies and technologies, in which nonhuman entities are simply resources or sinks for human development. Sometimes nonhuman nature is represented in the form of planetary boundaries or limits which must be treated with caution—possibly even *respected*—but dominant political imaginaries treat them as largely passive and malleable, a frontier for human mastery.

The closest consideration of nonhumans in mainstream political discourse occurs in conservation. Conservation discourses encompass so-called 'strong' conservation framings which seek to preserve and protect 'pristine nature' from human degradation, often focused on iconic species and landscapes, through to 'weak' conservation framings which transform nature into ecosystem services and seek to preserve those functions such as clean air and water that are useful to humans.[11] In both these accounts the line between humans and nonhumans is strictly maintained, both ontologically and in the form of conservation practices such as barriers and fences which separate the human from the natural, ostensibly for the protection of the latter but always in the ultimate interests of the needs and desires of the former. The drives to expand protected areas to '30 percent by 2030' or even set aside 'Half Earth' for protected nature are some of the most demanding visions for human–nonhuman relations that mainstream climate and environmental discourses can muster.[12] Turning such visions into reality would entail profound socio-political and economic transformations, certainly when compared to the absence of nonhuman species and landscapes from most ecopolitical imaginaries, but they do not exhaust the potential for radically re-envisioning human-nonhuman relations.

Climate fiction can provide a powerful alternative to this silencing of nonhuman nature in political imaginaries. Authors such as Amitav Ghosh, Margaret Atwood, Ursula Le Guin, Octavia E. Butler, Kim Stanley Robinson, Barbara Kingsolver, N. K. Jemisin, Paolo Bacigalupi, Maggie Gee, J. G. Ballard, Cormac McCarthy, and many others have explored richer and more diverse human–nonhuman relationships through their fiction. There is a prolific tradition of encounters with the Other in science fiction and fantasy, such as through the figures of the alien and the cyborg. In this they add imaginative depth to Mbembe's argument that a truly non-racial democracy would imply 'humankind ruling in common on behalf of a larger commons, which includes

[11] Büscher and Fletcher (2020); Duffy (2010); Ramutsindela (2007); Sinthumule (2016).

[12] Seehttps://iucn.org/resources/factsheet/conserving-least-30-planet-2030-what-should-count and https://eowilsonfoundation.org/what-is-the-half-earth-project/ (accessed 8 October 2024).

non-humans'.[13] Of course, not all fiction, even all climate or environmental fiction, necessarily decentres human or re-evaluates nonhuman perspectives. Indeed, the 'humanist' credentials of serious, realist literature that puts 'human life ... at the very centre'—as studied in *Human*ities faculties—are sometimes asserted as a strength in contrast to scientific or managerial discourses which foreground nonhuman agents and processes.[14] Queer theorists such as Nicole Seymour, however, have encouraged environmentalists to move beyond serious, realist, worthy stories, and consider the lessons to be learnt from absurd, ironic, inappropriate, frivolous, or playful texts.[15]

In this chapter I argue that Africanfuturist climate fiction makes a distinctively rich and original contribution to exploring nonhuman, more-than-human, and posthuman perspectives. In so doing I draw from and seek to build upon the work of scholars like Cajetan Iheka who has argued that African literature is fundamentally concerned with the 'interlinking of human and nonhuman lives', and that challenges to late capitalism 'must include a recognition of nonhumans as companions in a precarious world'.[16] None of the fiction examined here is about conservation or saving vulnerable species from extinction. There is a lot of tree-planting, it has to be said, but in most of these stories 'nature' is represented in ways that are complex, shifting, and anti-essentialist, amplifying the work of critical theorists in feminist and queer traditions. There are three parts to the story in this chapter.

First, the question of how to relate well and engage ethically with nonhuman others is prominent in most if not all the fiction examined in this book. Importantly, this is not posed as a counterfactual 'What if?' storyline, or an argument in favour of interspecies care, but rather proceeds from the starting assumption that we are all, always already bound up in mutually dependent relations with nonhuman others.[17] This gives Africanfuturist climate fiction a stronger more-than-human ontology and ethics when compared to mainstream climate fiction, certainly to pillars of the canon such as Ian McEwan's *Solar*, Kim Stanley Robinson's *Ministry for the Future*, and the films of Roland Emmerich such as *2012* and *The Day After Tomorrow*.[18] For Iheka, this is a result of 'the primacy of nature in African cultural systems'.[19]

[13] Mbembe (2019, p. 13; 2016, p. 335). On the alien and the cyborg in science fiction, see Haraway (2016); Heise (2008); Johns-Putra (2016, 2018, 2019); Thaler (2022); Trexler (2015).
[14] Nikoleris et al. (2017, p. 314).
[15] Seymour (2018).
[16] Iheka (2017, pp. 2, 5); see also Caminero-Santangelo (2014, pp. 55–74).
[17] Cudworth and Hobden (2022); Tschakert (2022); Whyte (2018).
[18] Johns-Putra (2019); Milkoreit (2017); Nikoleris et al. (2019).
[19] Iheka (2019, p. 203).

Second, in many of the Africanfuturist stories examined here, the new relationships that are imagined in climate-changed futures entail the emergence of quite radically new forms of human subjectivity. The transformation of human–nonhuman relations does not just mean the familiar environmentalist narrative of changing attitudes and values towards nature, but rather entails profound shifts in who or what humans are or might become. These stories can amplify and enrich theoretical work within postcolonial, feminist, and queer traditions on different modes of being human.[20]

Third, these changing relationships and subjectivities produce new visions of living with radically different others. Africanfuturist fiction is profoundly convivial, concerned with ways in which a more-than-human 'we' might rub along together in the future, but it is also insistently cognisant of the trouble that this will bring. This means seeing ourselves as 'intimate co-workers with the non-human in the mutual shaping of our present and future worlds'.[21] Going even further than this, living and working together also implies dying together.[22] Angry, predatory, homicidal species and environments are much more prominent in Africanfuturist climate fiction—including crocodiles, spiders, swordfish, vultures, dangerous waters, and inhospitable deserts—than the more familiar companion species of polar bears, butterflies, albatross, and trout that have become iconic in climate activism and fiction.[23] In exploring the world-views and relationships of unfamiliar nonhumans, Africanfuturist climate fiction can start to meet Ghosh's challenge 'of imaginatively restoring agency and voice to nonhumans'.[24]

Good Land Relations

The importance of restoring, rebuilding or transforming relations between human communities and 'the Land' in its broadest sense (including waters) is a central element of much indigenous philosophy and a core focus of indigenous climate fiction. These traditions have had a powerful impact on eco-philosophy and broader green political theory and imaginaries, even if they are frequently under-acknowledged in mainstream discourses. Good land relations are also a vital and prominent feature of many of the Africanfuturist

[20] Alberro (2024); Jackson (2020); Kafer (2013); Mbembe (2019); Seymour (2013); Wynter (2003).
[21] Hulme (2010, p. 274).
[22] Haraway (2016, p. 2).
[23] On these iconic species, see Duffy (2010); Liboiron (2021, pp. 105–6); Manzo (2010); Musila (2014, p. 152); Yusoff (2010).
[24] Ghosh (2021, p. 204).

stories examined here, in that they explore what Iheka calls an ethics and aesthetics of proximity to nonhuman nature.[25] An initial example of this comes in the utopian conclusion of Nnedi Okorafor's alien invasion novel, *Lagoon*. The President of Nigeria gives a speech in which he proclaims that alien help will enable the restoration of good land relations in Nigeria and beyond: 'the land would be pure and palm nuts, cocoa and other crops would grow as they never had before. Extinct creatures would return and new ones would appear. Nigeria would have much to give the world—and to show it.'[26] As Iheka suggests, 'planting or nurturing the soil' is both literal here and symbolic of broader national renewal.[27]

There are many other Africanfuturist stories in which plants, trees, and forests are represented in ways that stand for reciprocal relations of care and responsibility between humans and nonhuman others.[28] Luthando, a gay, black South African in Mackay's novel, *It Doesn't Have to Be This Way*, embarks on the first step of his journey to become an eco-activist when he attends a very white, very middle-class reforestation festival in the Platbos Forest with his white colleague, Malcolm, who is both in love with Luthando and obsessed with achieving a more accurate representation of trees and plants in the virtual world they are programming. Malcolm takes close-cropped photos of leaves and bark, documenting 'the gnarly shapes and idiosyncratic angles, the unexpected splits and breaks' of nature, whilst Luthando starts planting indigenous trees and doing guerrilla gardening on the slopes of Table Mountain. This can't prevent the climate catastrophe unfolding or mitigate the equally disastrous political, social, and technological responses that unfold, but at the conclusion of the novel, Milo, Luthando's adopted child, takes refuge in a cave and thinks about trees. 'Plants are the things to be', he decides in the final paragraph of the story.[29]

A visually powerful expression of this ecological sensibility comes at the end of Wanuri Kahiu's short film, *Pumzi*. The central character, Asha, receives a seed and a soil sample. She nurtures it with her own recycled water and smuggles it out of the compound, trekking across a wasteland to where she finds remnants of trees that, for Iheka, look like fig trees, 'reminiscent of

[25] Iheka (2017). There is an extensive literature on indigenous philosophy and land relations. Some of the most influential for my own research includes Davis et al. (2019); Davis and Todd (2017); Gilio-Whitaker (2019); Laduke (2004); Liboiron (2021); Maathai (2007; 2009); Ogude (2023); Ramutsindela (2007); Tamale (2020); Whyte (2018, 2020).

[26] Okorafor (2014, p. 279).

[27] Iheka (2017, p. 132).

[28] On trees and forests as ecological symbols more generally, see Fishel (2023); Iheka (2017); Maathai (2007); Maniates (2001); Strausz (2024).

[29] Mackay (2022).

Mugumo, the sacred tree in Kikuyu tradition.'[30] She eventually plants the seed and shelters it with her own body. The final shot of the film pans out from her prostrate figure—presumably her final resting place as she can't return to the compound—to show a vast forest in the desert. We hear the sound of cracking thunder then rain. Asha's act of care and nurturing for the tree that might sprout from the seed, and the existence of the wider forest, all appear to be expressions of hope, offering the prospect of a better future than that within the isolated and sealed dome.[31]

Iheka views this ending somewhat differently, arguing that the death of Asha limits the film's radical potential by excluding 'its female protagonist from the future that it envisions.'[32] In my view, while provocative and plausible, this reading depends on an anthropocentric ontology that Iheka elsewhere convincingly critiques. Rico describes the final scene as 'the transubstantiation of Asha's body into a tree' and suggests that some aspects of Asha will continue and live on through the wider environment and the rejuvenated arboreal ecology.[33] It is suggested that in the 'context of African folklore ... the transposition of Asha's body with a tree could imply a transferal of her spirit to the tree.'[34] I tend to agree, seeing this ending as a valorization of multispecies survival rather than a negation of the endurance of black women.

The short story 'Bittersweet are the Waters' also culminates in an act of rebellious, renegade seed-sowing. In the dried out Atlantic Basin of the far-future, Miyongi has been nurturing his grandmother's seedlings, an activity 'that would get him arrested for wasting water on non-humans for starts.'[35] But his plants prove invaluable when the insurrection against H2OCorp begins, led by the mysterious religious cult known as the Nymphomaniacs, who worship 'the tears of the Nymphs and the water of women' in strange ceremonies. Miyongi begins planting 'fairy-tale cacti' that germinate and shoot upwards almost instantly, 'tearing through steel hulls and bone in a geyser of bleeding metal.' Apple and pear cacti form impenetrable hundred-foot hedges against the corporate armies, and '[e]verywhere concrete bloomed.'[36] As the first rain in centuries falls it is the vitality of plant life that fertilizes revolution.

A tree and a seed are also central to the plot of Okorafor's novel, *The Book of Phoenix*. The central character, Phoenix, is being raised for scientific

[30] Iheka (2021, p. 37); Kahiu (2009). See also Rico (2017, pp. 90–1); Sides (2019, p. 114).
[31] Mackey (2018, p. 536); Nyawalo (2016); Omelsky (2014).
[32] Iheka (2021, p. 38). See also Moonsamy (2020, p. 189).
[33] Rico (2017, p. 90).
[34] Rico (2017, pp. 90–1).
[35] Diene (2022).
[36] Diene (2022).

research inside a tower in New York, where American scientists do experiments on enslaved 'speciMen', many taken against their will from Africa. The tower is built around a mighty tree, called the Backbone, which appears changed and enlivened by Phoenix's escape from the tower. Its 'great trunk was the diameter of two cars. Its rough rich brown bark was now covered with large sharp thorns. . . . If it wanted to, it could call its roots together, pull them out of the ground and walk away. . . . it had bloomed large fiery red flowers that grew high up.'[37] During Phoenix's escape she is caught, dies, and is then reborn. She is drawn back to the site of the tower by the 'forceful sagacious presence of the tree', and it gives her a seed, apparently of extra-terrestrial provenance. Phoenix returns the seed to Africa, eventually burying it under a shea tree in Wulugu, Northern Ghana, which brings 'great abundance' and fruitful flourishing to the trees in the region, as portrayed in Figure 4. Returning to New York, Phoenix sees a mob kill an angel and bring down the Backbone. At this, 'the chopping of a great tree, I sobbed with every part of my body. For everything.' This is a tree that had agency and emotions: of all the inhabitants of the tower 'the energy of its rage was the greatest'.[38] Catalysed by its destruction, Phoenix decides to enact a terrible retribution and unleashes an apocalypse causing the downfall of western civilization.

A forest plays a brief but important role in the sequel, *Who Fears Death*. Towards the end of the story, as the central character Onyesonwu follows her destiny and approaches her own foretold death, she travels in spirit form to a distant forest, a landscape strikingly foreign to the desert setting of the novel. She reflects that '[t]his place was too far to ever get to. But maybe someday it would not be. Maybe someday. Its vastness made the violence and hatred between the Okeke and Nuru seem small.'[39] In an otherwise bleak, violent, and destructive story, overshadowed throughout by the inevitability of Onyesonwu's death, this glimpse of a distant forest keeps a hope of a better future alive.[40] Onyesonwu recalls that '[w]hat I remember most about it was the deep sense of hope it placed in my heart. If a forest, a true vast forest, still existed someplace, even if it was very far away, then all would not end badly. It meant there was life outside the Great Book. It was like being blessed, cleansed.'[41]

[37] Okorafor (2015a, p. 37).
[38] Okorafor (2015a, pp. 38, 62, 197, 150).
[39] Okorafor (2018, p. 286).
[40] Dowdall (2013).
[41] Okorafor (2018, pp. 286–7).

Figure 4 Eric Battle, 'Reaper'

© 2015 Eric Battle. Previously published in Nnedi Okorafor, *The Book of Phoenix* (London: Hodder, 2016), p. 61.

Trees play an important symbolic role in many Africanfuturist short stories. One example is Vuyokazi Ngemntu's 'The Serpent's Handmaiden', set in KwaZulu-Natal, South Africa, in 2038, in which a father and daughter trek through dry scrubland in 48°C heat in order to fulfil a sacred prophecy. The only trees they see are thornbushes (the umnga tree) until they reach the mountains where the daughter has a transcendental experience 'kneeling under an olive tree'.[42] Her trance delivers the first rain for three years. The story closes with them heading home to plant orange trees, which haven't grown in the region for almost 20 years.

There is also a brief but significant role for a tree in Okorafor's novel *Lagoon*. After a violent confrontation between the alien Ayodele and Nigerian soldiers, she kills the soldiers then transforms their corpses into a 'plantain tree, heavy with unripe plantain'.[43] Later, after a riot in the area, '[t]he lawn looked as if it had been trampled by giants. Only the plantain tree was untouched'.[44] The tree seems to stand for the potential for a better future even in the aftermath of violence. Stories like this reflect the religious, philosophical, and mythical importance of trees—such as the baobab, palm, fig, mimosa, plantain, acacia, and iroko—and forests or sacred groves in many African cultures. Such trees have given cultural resonance to environmental campaigns such as Wangari Maathai's iconic tree-planting Green Belt Movement.[45] In these Africanfuturist stories, trees often appear to stand for stability, rootedness, care, interconnection, and an ecological sensibility, often in ways that have similarities to the symbolic invocation of trees and forests in many other global environmental movements and cultures.[46]

In contrast, trees are almost completely absent from Lauren Beukes' novel *Zoo City*. In an interview accompanying the book, Beukes reveals that '[t]he idea of the great South African novel is all about the journey into the interior, the wide expanses of the Karoo scrublands that expose the interior of the soul', whereas she wanted her novel to be about 'the sparking nerves, the guts, the pounding heart of the cityscape'.[47] Despite the urban setting, this novel provides one of the most vivid re-imaginations of human–nonhuman relations in contemporary Africanfuturist climate fiction. It is set in near-future Johannesburg, but there has been an 'ontological Shift' in global politics: criminals are

[42] Ngemntu (2022).
[43] Okorafor (2014, p. 137).
[44] Okorafor (2014, p. 176).
[45] Maathai (2007).
[46] Death (2016, pp. 112–14); Mbembe (2017, p. 180); Nixon (2011); Pahl (2018, p. 217); Rico (2017, p. 88); Tamale (2020, p. 87); Wright (2011, p. 241).
[47] Beukes (2010, p. 363).

now accompanied by animal familiars, bound together in a form of psychological symbiosis officially called 'Acquired Aposymbiotic Familiarism (AAF)'. Those afflicted are vernacularly known as 'zoos' or 'the Animalled'.[48] There are various theories about why this happened, including as a result of extreme forms of pollution and climatic change.[49] Inextricably linked to the human–animal couples are 'also mashavi—wandering spirits of the dead which attach to living humans', meaning that the novel 'stage[s] the human as inextricable from this landscape, its past, its present, and its nonhuman others'.[50] Indeed, those 'Animalled' cannot be physically distant from their acquired companion species without extreme physical and psychological pain: '[t]he pain is unbearable, you scream, you vomit, you say anything', one prisoner reports.[51] The death of the animal results in the obliteration of the human by the dark forces of 'the Undertow', meaning that individual humans are forced into close dependencies with individual animals, including lions, leopards, rats, and butterflies.[52] They must learn to live with their animals or face annihilation. According to Beukes, these close pairings between humans and nonhumans in the book are influenced by Southern African mythology, animism, and legends about animals and spirits, as well as Western authors such as Rudyard Kipling and Philip Pullman.[53]

The story follows the attempts by the central characters—Zinzi December and her animal, Sloth—to find ways of living together in inner city Johannesburg, and to deal with the debts of Zinzi's past: moral, criminal, and financial. The animals have their own personalities, agency, attributes, and symbolism. The reader meets and learns more about the Sloth, and Zinzi's lover's Mongoose, before any human character, and key protagonists—such as the Maltese (Mark) and the Stork (Amira)—are often referred to by their animal. There is a mutual dependence between Zinzi and Sloth that is unbreakable and that contrasts with the sequence of betrayals and fractured relationships in Zinzi's past: her estranged parents, her ex-lover Gio, her boss Vuyo, her current lover Benoît, and most tragically her brother Thando, whose death resulted in Zinzi becoming Animalled. Zinzi and Sloth bicker and complain, often resenting each other, but they also care for each other and save each other's lives on several occasions. Early in the story Sloth climbs on the narrator's back, 'fussing and shifting before finally he settles. I used to get impatient.

[48] Beukes (2010, p. 75).
[49] Beukes (2010, p. 183). See also Eatough (2017) and Ericson (2018).
[50] Ericson (2018, p. 30).
[51] Beukes (2010, p. 98).
[52] Beukes (2010, p. 245). See also Cloete (2015).
[53] Beukes (2010, pp. 3, 79).

But this has become an old routine for the pair of us.'[54] Sloth often grips tightly (and painfully) onto Zinzi's arms or shoulders to warn her of something or to act as her conscience. At one of Zinzi's lowest points, 'Sloth murmurs in my ear, soothing noises. He hates it when I shout.' When Sloth intervenes in a fight to save Zinzi, he almost gets his arm broken: 'Our connection is one-way. I can't feel his pain, but it's bad enough to see it in his face.'[55] Unlike all the other relationships they are still together at the end of the novel. Indeed, the experiences they have been through together seem to have taught Zinzi—hitherto, a selfish, sarcastic, wise-cracking addict—that self-sacrifice for those she loves might be the only way to pay back her debts and do something worthwhile with her life.[56]

The plot begins as a missing person narrative, with Zinzi employed for her special skills in finding lost things. Refugees, such as Zinzi's Congolese boyfriend Benoît, and lost people, such as his wife and children, appear and disappear throughout the novel. But the final third of the story turns into something more sinister: there are murders of homeless people and prostitutes where their animals are never found, and it turns out there is a serial killer targeting zoos for 'muti murders'.[57] A dream-reference to the film *Soylent Green* and a description of Johannesburg as 'self-cannibalizing' suggest an unpleasantly *familiar* ingredient in the traditional medicines being sold for vast sums.[58]

The connections between humans and their nonhuman environments are explored in depth and with insight in the novel. A background theme concerns the ways in which pollution, mining, environmental inequality, injustice, and human greed have produced a broken and violent society and environment. This deep connection between place, humans, and nonhumans is established in the first lines of the novel: 'Morning light the sulfur color of the mine dumps seeps across Johannesburg's skyline and sears through my window. My own personal bat signal.'[59] The mine dumps with their yellow dust provide the setting for several crucial scenes, along with the leafy suburbs, the dilapidated sewers, inhuman(e) slums and refugee shelters, seedy nightclubs, and the wildness of Zoo City/Hillbrow itself. The sheer number and variety of nonhuman metaphors and similes in the novel carries its own power and humour. Benoît has feet 'like knots of driftwood'; bedroom ropes hang like 'authentic Amazon

[54] Beukes (2010, p. 5).
[55] Beukes (2010, pp. 265, 316). See also Smith (2016, p. 354).
[56] Buxbaum (2017, p. 87); Propst (2017, p. 421); Stobie (2012).
[57] Beukes (2010, p. 308).
[58] Beukes (2010, pp. 290, 296).
[59] Beukes (2010, p. 3). See also Bethlehem (2015).

jungle vines'; Sloth has a 'mossy reek'; Benoît 'uncocoons himself from the bed'; humans are described as 'scrap rats', 'pack rats', 'lab rats', and 'tunnel rats'; her boss Vuyo 'has the air of the fox'; there are evictions by the Red Ants, the name of a real private security company specializing in evictions in South Africa; migrants are '[p]acked like tuna fish' and patrolled by 'sharks'; Odi and Zinzi are 'like chimps competing for dominance'; the sound of footsteps suggests 'a senile hippopotamus in fuzzy slippers'; Zinzi describes jealousy as a 'poison flower in my chest'; private security firms operate 'in the same way dogs piss on their territory' using 'boys built like bulldogs'; there are '[w]olf whistles and monkey whoops'; 'the war in the Congo is like an animal'; morning light feels 'like a mole digging its way into my skull'; drugs burn 'like speed cut with rat poison'; and at the novel's luridly violent conclusion the villain Odi Huron 'screams like a slaughterhouse pig in a PETA video'.[60]

The relations of more-than-human care explored in the novel are important and central to the story, even if the crime thriller plot twists and sensationalist violence sometimes seem to obscure them. In ecological terms it is significant that the core device of being Animalled remains particularized: Zinzi has to care for *her* Sloth, not all sloths. She develops no wider regard for animals or the environment more generally.[61] There is no romanticization of her relationship with Sloth: 'animals are still animals. And animals can be assholes, too.'[62] It is very clear that the ontological Shift has not led to a world in which human-nonhuman relations are being reconstructed on more ethical lines; in contrast the phenomenon is treated with sensationalism, as an exotic and exciting manifestation of violence, wildness, and eroticism. The pastiche form and style of the novel satirizes crime reportage, Tarantino-noir and true crime documentaries.[63] While the novel appears irreverent, sardonic, playful, and even perverse—all qualities Seymour associates with a queer aesthetic she terms 'bad environmentalism'—there is a wider social context of colonialism, capitalism, racism, misogyny, ecological destruction, and urban inequality ever-present just below the surface of the story.[64]

In my reading, the relationship between humans and their animals in *Zoo City* is primarily a symbolic one: Sloth reminds Zinzi of her crimes and debts, quite literally hanging around her neck. After all, 'nothing says guilty like a spirit critter at your side'.[65] She must learn to deal with them, live with them,

[60] Beukes (2010, pp. 3, 4, 6, 7, 41, 50, 51, 89, 94, 103, 151, 237, 242, 247, 251, 262, 268, 271, 351).
[61] Buxbaum (2017, pp. 86–7); Smith (2016, p. 357).
[62] Beukes (2010, p. 52, see also pp. 88–9, 120).
[63] Beukes (2010, pp. 134, 263).
[64] Seymour (2018, p. 4).
[65] Beukes (2010, p. 12). See also Hoad (2016); Stobie (2012, p. 377).

in an ethical way. When she tries to excuse her job sending scam emails, Benoît reminds her that '[w]e all have debts to pay!'[66] Those who refuse to live with the consequences of their deeds come violently undone in the end. This is exemplified by Odi Huron, whose crimes drive the plot to its blood-splattered conclusion as he uses muti magic to sever his link with his animal, the albino crocodile. Immediately after the successful magic ritual, Huron's crocodile uses its new-found freedom to attack and kill its erstwhile partner. In the post-Apartheid South African context, learning to live with the consequences of history and the crimes of the past remains one of the most pressing social and political questions.[67] The heavy symbolism of the white crocodile—Apartheid president P. W. Botha was known as *Die Groot Krokodil* (see Figure 5)—and Beukes' positioning of her story in relation to the 'great South African novel' suggests that the primary concern of the book is how to manage the relationship with the past.[68] In this, the animals are largely symbolic rather than agents of a new ecological politics. But, as Seymour argues, environmentalism can draw lessons from how such work employs 'dissident, often-denigrated affects and sensibilities to reflect critically on both our current moment and mainstream environmental art, activism, and discourse'.[69]

Going beyond the immediate South African context, it is worth thinking about the political implications of this story for global climate change debates. As will be discussed further in the next chapter, demands from the Global South for reparations to address climate and ecological debt are growing, going far beyond, quantitatively and qualitatively, the paltry sums pledged so far to the UNFCCC's Loss and Damage Fund. Long-term climate strategies and debates about energy transitions and green new deals must start to consider how ecological debts are to be repaid and what it might look like to relate ethically and convivially with more-than-human others in climate-changed futures. As Zinzi says, '[y]ou don't get to choose the ghosts that attach themselves to you'.[70]

The restoration of good Land relations—frequently represented in terms of care for trees—is an important theme in many Africanfuturist climate stories. Contrastingly, Beukes' *Zoo City* uses an intriguing plot device to explore human-nonhuman relations in a very modern urban African environment. In such ways these stories offer a route beyond simplistic appeals

[66] Beukes (2010, p. 265).
[67] Hoad (2016, p. 305).
[68] Beukes (2010, p. 363); Ericson (2018, p. 26); Stobie (2012, p. 377).
[69] Seymour (2018, p. 6).
[70] Beukes (2010, p. 16, see also p. 68).

Figure 5 Zapiro, 'PW Botha—Gross Violations'
© Zapiro 1996. Previously published in *The Sowetan*.

to 'care for Nature' or cultivate an ecological sensibility that critical theorists have so often derided. As Seymour insists, 'there must be a way to question "nature" and "environmentalism" while still caring for the non-human natural environment'.[71] Africanfuturist climate fiction provides some powerful examples of stories that help us think this through. Although some of these stories could be interpreted as being about humans learning to *care for* nonhuman nature, many stories take this a step further and begin to examine how human subjectivity and identity might itself be transformed. As Beukes has suggested, '[f]iction is how we grapple with ourselves'.[72] The next section considers this more closely.

Becoming Posthuman

Transformed relations with nonhuman Others could have profound implications for human subjectivity. As such, African climate fiction can play a role in offering 'alternative accounts of subjecthood' to those that dominate

[71] Seymour (2013, p. 10).
[72] Beukes (2017).

global political imaginaries.[73] Shape-shifting, or humans transforming into animals, is one way in which speculative fiction has often sought to represent posthuman subjectivity. The phenomenon of shape-shifting, positioned within a broader animist ontology, is a central element of many African philosophical and spiritual traditions and is reflected in the canonical literature of authors like Ngũgĩ wa Thiong'o, Amos Tutuola, Zakes Mda, and Ben Okri.[74] It is striking how many of the Africanfuturist climate fiction stories examined here involve central characters transforming into animals. One of the most obvious and most high-profile examples is in the Marvel film, *Black Panther*. T'Challa becomes the Black Panther through ingesting the heart-shaped herb, thereby acquiring the familiar superhero/super-animalesque powers of speed, strength, agility, and so on.[75] Another box-office example is Neill Blomkamp's sci-fi adventure film, *District 9*, in which insectoid alien refugees—known derogatively as 'Prawns'—are confined to an internment camp after their moribund spaceship is stuck over Johannesburg. In the film the central character, a human bureaucrat called Wikus van de Merwe, is exposed to an alien substance and begins to transform into a Prawn.[76] Notwithstanding the very contrasting treatments of these two transformations—one noble and heroic, the other portrayed as degrading and pitiable—they both illustrate the prominence of posthuman and shape-shifting tropes in Africanfuturist stories.

Nnedi Okorafor's novels feature several shape-shifting characters. Onyesonwu in *Who Fears Death* is an 'Eshu' or shapeshifter. Her first transformation is into a sparrow. It is unintentional and terrifying, and she ends up naked and 'stuck high in the giant iroko tree that grew in the centre of town'.[77] As she learns more about the process of transformation she usually chooses to become a vulture, becoming able to feel 'the vulture and the sparrow just below my skin'. The sensory transformations are powerfully rendered in the novel in a way that is difficult for other mediums. 'I flew high. My sense of touch was less, for my flesh was protected by feathers. But I saw all. My hearing was so sharp that I could hear the land breathing.'[78] Near the end of the story she transforms into a Kponyungo dragon, with her mother, in order to fly together to a secluded forest. She can also enter and pass through the spirit world as her training in the Mystic Points progresses.

[73] Sabaratnam (2011, p. 785).
[74] Arseneault (2019); Caminero-Santangelo (2014); Hugo (2017); Iheka (2017); Mbembe (2017; 2019); Nyamnjoh (2017); Okorafor (2009); Olaoluwa (2018); Slaymaker (2001, p. 136); Wright (2011).
[75] Coogler (2018).
[76] Blomkamp (2009).
[77] Okorafor (2018, p. 15).
[78] Okorafor (2018, pp. 57–8).

This is dangerous: there are blue spirit trees that try to strangle her and the baleful red eye of her rapist biological father, Daib, watches her and plots her destruction. He also has the power to shape-shift and transforms into a bat. As Onyesonwu grows in strength and power her ability to enter and draw on the wilderness improves, using it to transform, travel, become invisible and heal or destroy. However, there is a danger in shape-shifting: '[t]o remain a vulture too long always left me feeling detached from what I could only define as being a human.... There is always a price for changing.'[79] The twist at the very end of the book is that the prophesized ending, Onyesonwu's death, might not be the only possibility. An alternative ending is suggested, where she transforms into a Kponyungo dragon, torches her captors, and then 'blew into the sky like a shooting star, ready to return home', heading for the iroko trees.[80]

More-than-human ecologies also take centre stage in Okorafor's *Lagoon*. Act I opens with a chapter written from the perspective of a swordfish, Act II from a tarantula, and Act III from a bat's perspective. The story is narrated by Udide Okwanka, 'the story weaver, the Great Spider', who has the first and last word and spins a story rooted 'in the dirt, the mud, the earth, in the fond memory of the soily cosmos.'[81] The book itself is dedicated 'to the diverse and dynamic people of Lagos, Nigeria—animals, plant and spirit.'[82] The story concerns the arrival of shape-shifting aliens in Lagos, a plot that emerged from Okorafor's reaction to South African film *District 9*. She credits it for 'intriguing and pissing me off so much that I started daydreaming about what aliens would do in Nigeria.'[83] The aliens interact with various human, nonhuman and mythical entities, including the Bone Collector, a road monster who angrily consumes humans and vehicles, as well as transforming the human population of Lagos in profound ways. Significantly the aliens make contact first with the nonhuman occupants of the waters, only then reaching out to humans: '[w]e landed in your waters and have been communicating with other people there and they've been good to us. Now we want *your* help.'[84]

The three central human characters are Adaora, Agu, and Anthony. They are the first to be contacted by the aliens—in the form of the alien ambassador, whom Adaora names Ayodele—and it emerges that they all have special

[79] Okorafor (2018, p. 82).
[80] Okorafor (2018, p. 385).
[81] Okorafor (2014, pp. 194, 228).
[82] Okorafor (2014, p. v).
[83] Okorafor (2014, p. 301).
[84] Okorafor (2014, p. 37). See also Hugo (2017); Jue (2017); Mackey (2018); O'Connell (2016, pp. 309–10).

powers and attributes. Adaora is a successful scientist, a marine biologist with a love of the sea. We find out later that she was born with webbed feet and hands and could swim without being taught. She has the power to create force-fields and can use them to walk on water. As the novel reaches its conclusion the aliens help her transform into a mermaid, and her physical transformation into a creature with gills and a tail confirms her suspicions (and the allegations of her husband) that she has always been a marine witch.

Agu—whose name means leopard—is a Nigerian soldier with superhuman strength. His strength 'lived in him like the roots of a thousand-year-old tree in its plot of land.'[85] Anthony is a successful Ghanaian rapper who can weave words and music into waves of power: '[w]hen he performed, he spun words as a spider spins its web.'[86] These three central characters are close to the nonhuman world—elements, spirits, and animals—and this is apparently why they have been chosen for the first human contact. Their abilities pre-exist the alien arrival but the visitors, who can manipulate matter and energy, help them in their transformations. Shape-shifting aliens seem to offer a lesson to Lagos' inhabitants that humans can be otherwise, more than they might imagine. As Ayodele tells them, '[w]e are change.'[87] The potential implications of this are not lost on the members of an LGBT campus group in Lagos, who call themselves the Black Nexus. When they see the video of Ayodele changing form they immediately recognize the power of the message: 'that's a person changing into another person! Would've been better it if changed from a woman to a man but this will do. We could have some fun sending this around. . . . The Black Nexus can come out of secrecy for *this*. Who better to understand than a shapeshifter?'[88]

As the novel reaches its conclusion Ayodele sacrifices herself to heal and transform all the inhabitants of Lagos. She is viciously beaten (again) by soldiers after they return from the meeting between the President and the Elders and this time, instead of responding with violence, she submits to the attack. Her last words to Adaora are: '[y]ou people need help on the outside but also within . . . I will go within. . .' Ayodele turns herself into a white mist which expands until everyone in Lagos inhales it, transforming them all. 'Ayodele was gone. Ayodele was here. "Lagos will never be the same", Adaora said.'[89] The transformations enabled by the aliens are extensive and profound, but

[85] Okorafor (2014, p. 174).
[86] Okorafor (2014, p. 165).
[87] Okorafor (2014, pp. 39, 112).
[88] Okorafor (2014, pp. 73–4).
[89] Okorafor (2014, pp. 268–70).

not total. Even the President, for example, despite everything he has been through and has seen Adaora, Agu, Anthony do, and his intention to rely on them as experts in the future, '*still* felt Agu, Adaora and Anthony were witches. Good witches, but witches nonetheless. Old outdated ways of thinking don't die easily, and sometimes they don't die at all.'[90] Shape-shifters are mistrusted and often treated violently in Okorafor's novels, but in *Who Fears Death* and *Lagoon*, as in *Black Panther*, they represent a posthuman future in which humanity can transcend its limitations and help forge a better world.

Several Africanfuturist short stories revolve around transformations in human–nonhuman relationships and subjectivity. At the culmination of Jessica Wilson's 'The Drought', a young South African girl, Karipi, several centuries in the future, goes through a transcendental rite-of-passage where she experiences what it feels like to be 'the last fish-eagle to be born' and a 'river desperately trying to rid itself of poisons'.[91] Caught in a rock pool in a storm and near to drowning, she swims downwards in a trance until she meets 'an enormous moon-coloured ray'. This turns out to be an embodiment of her grandmother. On they swim, an 'ancient ray and young girl, human-fish and fish-human', until Karipi observes one small fish repairing a bleached coral reef and 'she knew what she had to do'. She leaves her trance and the ocean, returning home to find her grandmother has sacrificed her life to save her granddaughter. Despite her sadness she knows she has 'found my task, my calling' and feels in touch with 'every living being throughout the universe'.[92]

Ben Okri has insisted that in rising to the challenges of climate change, 'we need to become a new people'.[93] Through the representation of shape-shifters and transcendental experiences of nonhuman perspectives these stories illustrate some radical ways to imagine posthuman futures. In this respect speculative fiction has far greater imaginative power than so-called realist writing, or nonfictional political texts, to represent more-than-human and posthuman politics.[94] In-so-doing new political possibilities can be opened up, although of course there are also dangers involved.

One risk concerns the racialized histories of the use of animalistic language and imagery in representations of Africans. As Mbembe famously asserted, 'discourse on Africa is almost always deployed in the framework

[90] Okorafor (2014, p. 278).
[91] Wilson (2019, p. 37).
[92] Wilson (2019, pp. 37–9).
[93] Okri (2023, p. 190).
[94] Arseneault (2019); Chakrabarty (2012); Iheka (2017, p. 17).

(or on the fringes) of a meta-text about the animal—to be exact about the beast'.[95] There is a long and deeply problematic history of imposed associations between animals and Africans, Blacks and other racialized groups. For this reason, Marvel decided that the character of M'Baku in the *Black Panther* film would not be referred to by the name of the character in the comic-book version, 'Man-Ape', nor would he wear a gorilla costume. Executive Producer Nate Moore reflected that '[h]aving a black character dress up as an ape, I think there's a lot of racial implications that don't sit well, if done wrong. . . . It's a line I think we're walking, and hopefully walking successfully.'[96] This is a line that has to be walked carefully in other stories examined here too and it is easy to understand a reluctance to invoke posthuman and transhuman perspectives by writers from communities who have often been denied full membership of 'Humanity'.[97] Yet, for some of those working in the traditions of black political theory, it is precisely this dehumanizing history that energizes the importance of questioning and subverting the categories of 'Man' and 'Mankind'. Zakiyyah Iman Jackson argues persuasively that discourses of animalization must be challenged (especially by Critical Black Studies) on two fronts: rejecting 'animalising discourse that is directed primarily at people of African descent, and animalising discourse that reproduces the abject abstraction of "the animal" more generally because such an abstraction is not an empirical reality but a metaphysical technology of bio/necropolitics applied to life arbitrarily'.[98] Drawing on rich literary and philosophical traditions of animism, the stories examined in this section put the coherence of anthropocentrism into question by portraying blurred and shifting lines between humans and nonhumans. By transgressing human–nonhuman binaries it is also possible to destabilize the hierarchies through which diverse other subjects have been denied the status of full humanity—women, transgender, intersex, queer, disabled, neurodiverse, racialized, and others—and arrive at more capacious and inclusive imagined subjectivities for climate-changed futures. However, complicating and defamiliarizing arbitrary distinctions between human and nonhuman does not automatically produce a politics of consensual, pluriversal egalitarianism. Living alongside more-than-human Others can be dangerous, troubling, and uncomfortable, as the next section explores.

[95] Mbembe (2001, p. 1). See also Arseneault (2019, p. 232); Ghosh (2021, p. 195); Iheka (2017, p. 160); Jackson (2020); Moonsamy (2020); Tamale (2020).

[96] Breznican (2017).

[97] Mbembe (2019, p. 164); Slaymaker (2001).

[98] Jackson (2020, p. 15). See also Davis et al. (2019); Jackson (2020); Mbembe (2019); McKittrick (2015); Weheliye (2014); Wynter (1987, 2003).

Living with the Radical Other

There is a risk in some representations of more-than-human futures that they tend towards apolitical fantasies, in which power relations and violence have been magicked away, transformed into a web of cooperative and care-full relations. Inspired by Iheka's work, which focuses on violence, struggle, and representations of the 'brutalities of war' in more-than-human African ecologies, the third part of this chapter focuses on the pain, difficulty, and trouble of living with others.[99] It is here that Africanfuturist climate fiction makes one of its strongest and most distinctive contributions to imagining new ecological futures. In these stories, the struggle to find ways to live alongside each other continues, often in some of the most foreign or hostile environments imaginable and alongside some of the most radical Others. In this, Africanfuturist climate fiction continues a rich tradition in Afrofuturism of imagining forms of co-existence between humans and others who appear (at least initially) repulsive or abhorrent.[100] The most famous example is Octavia Butler's iconic short story 'Bloodchild', in which the male human Gan is required to become a host for the eggs of the insectoid T'Gatoi (a member of the Tlic species), a relationship that is not (just) one of master-slave or parasite-host, but one of affection and intimacy.[101] The politics of living alongside the radical Other—*convivial* politics—is explored here in two ways. First, through Africanist renderings of familiar eco-dystopian narratives, and second, through the ongoing power relations and violence of (some) more-than-human encounters.

Eco-dystopian narratives are very familiar: they tell the story of human suffering and resilience in the face of a hostile and changing environment. Doris Lessing's *Mara and Dann* novels are in this respect fairly conventional environmental dystopias. In the first ten pages of *Mara and Dann* the children experience drought, extreme thirst, and then a flash flood. 'Mara knew that animals were dying everywhere because of the dryness', and she is quickly made aware that many of the animals they meet—water dragons, scorpions, spiders, large insects—are hungry and dangerous.[102] The shifting climate means that everything is changing: rivers, forests, 'and insects, even scorpions, changed their natures'. In this context Mara and her little brother become 'two hunted and hunting creatures', and the absence of the 'vast forests' that used to cover

[99] Iheka (2017, p. 57). See also Haraway (2016, p. 1); Liboiron (2021, p. 110); Tschakert (2022, p. 289).

[100] Bryce (2019); Smith (2012, pp. 151–2).

[101] Butler (2005). See discussion in Bollinger (2007); Helford (1994); Jackson (2020).

[102] Lessing (1999, p. 12).

central Ifrik are a symbol of the dystopian present.[103] Of all the Africanfuturist stories in this book, Lessing's are some of the most apparently anthropocentric and humanist. When humans are described as 'like animals' it is usually pejorative, and Becket critiques Lessing's ecological politics in these novels for their failure to fully interrogate 'human centrality' or 'offer a critique of the complexities of human authority in relation to environmental catastrophe'.[104]

I read Lessing's ecology as more complex than this, though. There are, at key moments, glimpses of a less anthropocentric politics. When Mara experiences drought in Chelops 'the flowers seemed to Mara to be emitting a high, almost audible scream for help, because she was remembering, and identifying so strongly with a longing for rain'.[105] Later, when they have reached a safer climate in the North, Mara wonders why she feels compelled to keep moving: 'Mara looked back down her life . . . and seemed to herself like a scurrying little beetle.'[106] There is a moment in the sequel, again striking for being so rare, where Dann reflects that '[i]t seems we can be lizards in sands and then take to water like birds. All our early lives Mara and I were dying of drought and then we became water people.'[107] These brief, limited moments that hint at a more-than-human perspective need to be placed in the context of an epic narrative that is centrally concerned with the question of what it means to be human—to tell stories, to remember the past, to use technology, and to preserve fragments of culture and literature—in a context where life has been reduced to the daily struggle for survival in a hostile environment.[108]

The sequel, *The Story of General Dann and Mara's Daughter, Griot and the Snow Dog*, takes these questions further, and in ways that existing critical analysis of Lessing's eco-politics do not discuss.[109] For the first time in the *Mara and Dann* novels there is an important nonhuman character: Ruff, the snow dog. Dann saves the pup from the marshes and the dog saves Dann's life on several occasions. There is a depth and intensity of emotional relationship between the two that is touching, especially in the context of Dann's difficulty in relating emotionally to human characters.[110] There are also significant moments of encounter with packs of other snow dogs, who are making their own migrations for their own, unknowable, nonhuman reasons. On the one hand, Dann's

[103] Lessing (1999, pp. 67, 104, 217–18).
[104] Lessing (1999, p. 206); Becket (2009, p. 135).
[105] Lessing (1999, p. 143).
[106] Lessing (1999, p. 305).
[107] Lessing (2006, p. 214).
[108] Kuznetski (2021); Sargeant (2016); Walder (2008); Watkins (2012).
[109] Hanson (2016); Sargeant (2016); Watkins (2012).
[110] Lessing (2006, pp. 27–33, 84–6, 257, 270).

intense emotional bond with his dog is fairly familiar, indeed domesticated, and will be recognizable to many readers who have had a close canine companion in their lives.[111] While Ruff seems to be an intelligent and sensitive animal we never get his perspective or a meaningful sense of Ruff's agency, even when Ruff is made seriously ill by the conflicting needs of Dann and Tamar, both of whom he loves. On the other hand, in death, Ruff's symbolic role is amplified. His passing is introduced obliquely, in a manner closely reminiscent of the news of Mara's death at the start of the book. Moreover, Mara and Ruff are the only major characters in these epic novels whose deaths are mourned and commemorated. Dann and his companions build a monument to Ruff, a striking moment in stories whose central theme is the unreliability of memory and the impossibility of anything surviving over time.[112] As the epic story closes, Griot sings the tales of Mara and Dann to the foxes and birds on the high moors, sometimes wordlessly. He realizes, with some concern, that 'Ruff could have howled like that'.[113] In these brief glimpses of a wider ecological web that encompasses the human and nonhuman characters, perhaps we can see Lessing allowing the possibility that people—and their stories and memories—will find ways to live alongside nonhumans in the changing climates of the far future.

There are many other examples of Africanfuturist climate fiction that imagine a dystopian future in a hostile more-than-human world. In Kahiu's short film *Pumzi* humans huddle in compounds to protect themselves from the irradiated East African desert outside.[114] In Okungbowa's short story, 'Dune Song', the inhabitants of a far-future settlement, Isiuwa, live beyond barricades to keep the desert out, and 'scuttle along, dressed in cloth wrapped to battle every iteration of dust-laden wind'.[115] In Okorafor's novel *Noor*, a terrifying sandstorm called the Red Eye scours the harsh desert landscapes of northern Nigeria.[116] In Tlotlo Tsamaase's short story 'Eclipse our Sins', set in the far-future land of Kang, humans hide inside dystopian cities and breathe through filters to protect themselves from a dangerous world. Mama Earth itself, and the air, sun, seas, and diseases, have become implacably hostile to human existence. This is their revenge for human sins: 'carbon emission, racism, oil spills, sexism, deforestation, misogynism, xenophobia, murder'.[117] The narrator

[111] Haraway (2016, p. 105).
[112] Lessing (2006, pp. 270, 274).
[113] Lessing (2006, p. 276).
[114] Kahiu (2009).
[115] Okungbowa (2019).
[116] Okorafor (2021).
[117] Tsamaase (2019).

watches her family members die in anguish from their sins and those of their species: 'Mmê Earth, You used to be so healthy for us until we destroyed You. I understand now why You want to purge us from Your womb. But it is unfair. How come we are the ones to suffer for the before-generation's desires that smoked our future?' She concludes 'We have become foreigners to Earth, and it has turned xenophobic' and the only possible way out is cataclysm: '[s]ea-rise will be our baptism'.[118] This vengeful Earth—with some similarities to N. K. Jemisin's *Broken Earth* series—is an example of what I have described elsewhere as 'wild necropolitics', in which the chaotic violence of 'nature' or 'the wilderness' is positioned as a central feature of climate-changed futures.[119]

Another example is Osahon Ize-Iyamu's short story 'More Sea than Tar', about a family struggling to survive in a flooded Nigerian slum-world. The waters are polluted and diseased, full of 'animals merged with glass and plastic, mosquitos breeding, insects mutated by industrial waste and chemical reactions'.[120] On an entrepreneurial excursion they spot 'a crocodile dancing in the depths . . . covered in greasy film and toxic waste. Rubbish that has . . . congealed and grafted to the animal's skin—oh my goodness—like an infection sewn and healed onto the body'. Trying to find work, a father and son take a job requiring them to supply the corpse of a crocodile. They nearly die in the attempt, and as the monster rears out of the water 'pieces of glass stab into the beast's head and broken cans stick out of its body like piercings. Nylon bags are draped over this creature that is trying to eat me, and I scream and scream.' They succeed in killing the crocodile but are double-crossed by rivals and the story ends in tragedy. Even in this nightmarish landscape, amidst the violent deaths of his father and brother, the protagonist resolves to keep going and to try to pass on a message of a 'hopeful future'. The final line reads: 'I know which direction to go in. I'll find my way home.'[121]

The main challenge in these stories is how to live Otherwise in an environment that is troubling, uncomfortable, unsettling, and inhospitable. This is a challenge that is explored at length in Okorafor's novel *Lagoon*. Many readings of *Lagoon* emphasize the utopian conclusion and the promise of a new age of peace, plenty, and tolerance that the aliens and the President seem to offer.[122] These readings underestimate the violence, pain, and mayhem that is so central in the story. Transformation and change is not easy or straightforward,

[118] Tsamaase (2019).
[119] Death (2022a). See also Thaler (2022, p. 142).
[120] Ize-Iyamu (2019).
[121] Ize-Iyamu (2019).
[122] Eatough (2017); Hugo (2017); Jue (2017); Kotecki (2020); Mackey (2018).

even with alien assistance. Welcoming and living hospitably with the Other is not comfortable. This is insinuated in the description of the swordfish with which the story begins: 'She slices through the water, imagining herself a deadly beam of black light. The current parts against her sleek, smooth skin. If any fish gets in her way, she will spear it and keep right on going. She is on a mission. She is angry. She will succeed and then they [humans] will leave for good. They brought the stench of dryness, then they brought the noise and made the world bleed black ooze that left poison rainbows on the water's surface.'[123] There is no reconciliation between humans and the swordfish in *Lagoon*: she hated humans before the aliens arrived and the aliens help her transform into something even more deadly. She tries to kill the main human characters when they are travelling to meet the aliens and almost succeeds. At the end of the novel there is no truce, instead 'she swims away. South. She swims out to sea, to see what she can see.'[124]

Violence is integral to *Lagoon* on many levels. The story begins with the swordfish breaking an oil pipeline. In the aftermath she hears/feels a huge explosion: 'MOOM!' This, we learn, is the sound of the aliens sending out a sonic blast to 'read' Lagos. While it merely surprises or stuns creatures the size of the swordfish or humans, '[s]everal smaller fish, jellyfish, even crabs, float, belly up or dismembered. Many of the smaller creatures have probably simply been obliterated.'[125] This happens again with the second sonic boom: 'birds, insects and bats fell to the ground, dogs barked, cats hid, lizards scurried, several forms of bacteria died and others germinated. The noise this time was so profound that many of the weaker multi-cellular organisms in parts of the ocean closest to the source were obliterated.'[126] In a novel concerned with the diversity of life in all its forms, this repeated annihilation of life should not be passed over lightly. Contact with the Other is dangerous and violent. Transformative change is destructive and painful, and not just for nonhuman life.

The alien arrival transforms the waters of the lagoon. They are 'now so clean that a cup of its salty-sweet goodness will heal the worst human illnesses and cause a hundred more illnesses not yet known to humankind. It is more alive than it has been in centuries and it is teeming with aliens and monsters.'[127] The waters are now hostile to humans: the sea causes allergic reactions when

[123] Okorafor (2014, p. 3).
[124] Okorafor (2014, p. 290). See also O'Connell (2016, pp. 306–7).
[125] Okorafor (2014, p. 4).
[126] Okorafor (2014, p. 114, see also pp. 128–9).
[127] Okorafor (2014, p. 6).

the main characters journey out to meet the alien Elders and the creatures of the lagoon try to kill them. When Adaora asks Ayodele what the aliens have done to the water, she replies '[n]othing that didn't want to be done'. As the narrator, Udide warns, '[t]he sea always takes more than it gives', and 'there may well be a sign on all Lagos beaches that reads: "Here There Be Monsters". This has always been the truth, but today it is truer.'[128] As the President's party make their way back to land they learn that a giant octopus has killed ten of the soldiers waiting to meet them.

The aliens try to help resolve some of these conflicts, and in difficult circumstances they work hard to communicate and connect with humans. Their aims are peaceful and collaborative. Ayodele tells them, '[w]e are here to nurture your world.'[129] She tells the President directly, '[w]e do not want to rule, colonise, conquer or take. We just want a home.'[130] The aliens will give the nonhuman and human occupants of Lagos what they want in return. Yet they remain very clearly still alien, foreign, uncanny. Adaora's first reaction to the alien she calls Ayodele is that she appears 'both attractive and repellent . . . it addled Adaora's senses. . . . She had piercing brown eyes that gave Adaora the same creepy feeling as when she looked at a large black spider. Her mannerisms were too calm, fluid and . . . alien.'[131] The sound Ayodele makes when changing form is 'the sound of metal balls on glass' and it sounds '[a]bsolutely foreign. So foreign, that hearing it made her feel like falling to the floor.' Ayodele tells Adaora's children why she chooses to appear in human form: '[h]umans have a hard time relating to that which does not resemble them. It's your greatest flaw.'[132]

Relations between humans and aliens descend into violence on numerous occasions. Ayodele bears the brunt. After her first address to Lagosians she is shot by soldiers. She feels shock and pain: 'her left eye had dissolved to nothing but the look in her still intact right eye was one of pure hatred'. In retaliation she kills the soldiers: '[w]here the soldiers had stood, heaps of raw meat wriggled and then became still'. Ayodele is unharmed but her outlook has changed. She is now 'scowling at Adaora' and '[t]he warm, curious, light-hearted being that Ayodele had been was gone. The eyes Adaora looked into now were those of an angry, bitter old woman.' Her message of peace and love has been rejected, and the ambassador is tempted to refuse her role: '"I hate humans", Ayodele

[128] Okorafor (2014, pp. 248, 228).
[129] Okorafor (2014, p. 113).
[130] Okorafor (2014, p. 220).
[131] Okorafor (2014, p. 17, see also p. 143).
[132] Okorafor (2014, pp. 30–1, 67).

said.... "I want nothing to do with you", Ayodele continued. "Any of you."' She then decides to change herself into a tiny monkey 'because she'd decided that she no longer wanted to be a human being.'[133] When she is eventually beaten into oblivion by soldiers the reader is guided to interpret it as an act of restraint, love, and sacrifice by a powerful alien force. Yet as Nedine Moonsamy observes, what we are actually shown, in grisly detail, is extreme violence being inflicted on 'the body of a young black woman.'[134]

The arrival of the aliens also intensifies human relations with other, older, 'greater beings of the earth, soil, sea, lagoon and land', and these are also dangerous to live alongside.[135] There is a stretch of the Lagos-Benin Expressway that 'has named itself the bone collector. It mostly collects human bones, and the bones of human vehicles.' It swallows the bodies of those killed on the road, until an alien woman speaks to it and offers herself. After her sacrifice, '[t]he road shuddered, the road stretched but in the end, the road was satisfied. It laid itself back down, and became still. And when it was done, I heard the relieved sigh of millions of ghosts.'[136]

Even in the new world that the aliens have helped create it is clear that human–nonhuman relations will not be smooth. The bat whose perspective opens 'Act III: Symbiosis' is hit by the President's plane returning to Nigeria. The pilot remains unaware that they have 'just killed the most enlightened bat on earth.'[137] Adaora—after all she has been through—continues to think like a scientist, albeit one with a greater awareness of nonhuman agency. She plans to 'collect samples and creatures (at least the ones who would allow themselves to be collected)' and study them to understand what has happened to the waters. The final words of the novel are left to Udide Okwanka, spider the artist. Concerned at the hostile reaction of other countries to what is happening in Lagos, he resolves that '[f]or the first time since the birth of Lagos, my glorious city, I will pause in my storytelling. I will leave my web. I become part of the story. I will join my people. And we spiders play dirty.'[138]

The title of this final chapter, 'Spider the Artist', echoes Okorafor's short story with the same name. Also set in Nigeria, it is a very different story about a relationship between a woman, Eme, and a spider-like security machine designed to guard the oil pipeline but which is attracted to Eme's guitar playing. 'Its lovely eyes would glow with joy. Could a robot feel joy? I believed intelligent

[133] Okorafor (2014, pp. 136, 138, 140, 153).
[134] Moonsamy (2020, p. 188).
[135] Okorafor (2014, p. 120).
[136] Okorafor (2014, pp. 120, 208, see also pp. 171–2, 202–9).
[137] Okorafor (2014, p. 225).
[138] Okorafor (2014, pp. 280, 292–3).

ones like this could.'[139] It spins itself a guitar and starts to play, and 'its music deepened to something so complex and beautiful that I was reduced to tears of joy, awe, ecstasy'. Eme names her droid Udide Okwanka. But here, even more starkly than in *Lagoon*, there is no happy ending, no utopian resolution in which humans and nonhumans learn to live alongside in peace. The spider droids—known as Zombies—turn on the people, causing an explosion that kills hundreds of children and adults. Eme is saved by the droid, but a wider war between humans and the Zombies loom. The story concludes with a warning: 'Pray that Udide and I can convince man and droid to call a truce, otherwise the delta will keep rolling in blood, metal and flames. You know what else? You should also pray that these Zombies don't build themselves some fins and travel across the ocean.'[140] This terrible future comes to pass in Okorafor's novel, *The Book of Phoenix*.[141] Living with dangerous Others is not easy. It is painful and often violent.

This is the challenge set by the representation of more-than-human futures in these Africanfuturist climate novels. How can we live alongside Others, in difficult and violent circumstances? How can we do so in ways that are worthwhile, ethical, just, sustainable, responsible, and hospitable? As Lauren Beukes reflects on her experiences doing research in inner-city Johannesburg, 'people find ways to live with ghosts and that's why we're here, because despite the horror stories . . . Hillbrow is somewhere people live'.[142] For Iheka, the condition of living alongside human and nonhuman Others is best described as convivial, 'a mutual imbrication, a consequence of living together and sharing physical space' that can be contentious and violent, full of frictions, and tensions.[143] This concept is worth exploring a little further at the close of this chapter.

The foundational theorists of conviviality are Ivan Illich and Paul Gilroy, whose arguments have sometimes been interpreted as rather rosy images of multicultural tolerance and interdependence.[144] Sometimes conviviality seems to mean just 'getting along well' with others. For Bram Büscher and Robert Fletcher, convivial conservation is their aspiration, a democratic, postcapitalist, 'lived relationship to multidimensional (human and nonhuman) natures'.[145] In contrast to these accounts of conviviality, Iheka rightly emphasizes its conflictual and troubled aspects.[146] In this he draws on Mbembe

[139] Okorafor (2011).
[140] Okorafor (2011).
[141] Okorafor (2015a, p. 118).
[142] Beukes (2010, p. 368).
[143] Iheka (2021, p. 153).
[144] Illich (1973); Gilroy (2005). See also Nyamnjoh (2017).
[145] Büscher and Fletcher (2020, p. 64). See also Alberro (2024, pp. 232–3).
[146] Iheka (2021).

whose interpretation of conviviality as the postcolonial condition is significant but often misinterpreted. For Mbembe, conviviality is a forced sharing of 'the same living space' by the powerful and the colonized, in which inescapable 'familiarity and domesticity' lead to 'mutual zombification' and a situation where 'each has robbed the other of vitality and left both impotent'.[147] As a result of the domestication of colonized by colonizer, Mbembe stresses, neither 'emerge from this circle unharmed'.[148] The postcolonial context of absurdity, banality and the carnivalesque is far from an ideal form of social relation, but a fetishized, paradoxical, mutually interdependent questioning and reconfirmation of power.

Much of Mbembe's work since the publication of *On the Postcolony* has been a continued search for ways 'to exit the Fanonian cul-de-sac' of violent struggle and death as a response to forced colonial conviviality.[149] In *Necropolitics* he explores the question of whether, as human and nonhuman, we can 'found a relation with others based on the reciprocal recognition of our common vulnerability and finitude?'[150] This will entail, he insists, living 'in-common', based on an 'idea of a world that is the only one we have and that, to be sustainable, must be shared by all those with rights to it, all species taken together'.[151] The challenge, as he sees it, is to bring together 'the agents and milieus of life—water, air, dust, microbes, termites, bees, insects' in a 'democracy of species'.[152] Yet every democracy 'bears the colony within it' and biopolitical/necropolitical techniques are ineradicable elements of contemporary planetary entanglements.[153] This does not evacuate the concepts of choice, decision, and agency. We can attempt to choose how we live and die with others, albeit not in circumstances of our own choosing.

Fictional stories are powerful tools in representing and interrogating some of these choices. The multiple representations of the deaths of central characters—human and nonhuman—in these stories, including Asha in *Pumzi*, Ayodele in *Lagoon*, Huron in *Zoo City*, Mara and Ruff in *The Story of General Dann and Mara's Daughter, Griot and the Snow Dog*, Nata and Tasé in 'Dune Song', and Onya, Mwita, and Luyu in *Who Fears Death*, all suggest there are lessons about dying well (and badly) in more-than-human multispecies entanglements. Francis Nyamnjoh argues that a central tenet of much African philosophy is that '[d]eath is a form of circulation and not a matter of

147 Mbembe (2001, p. 104).
148 Mbembe (2001, p. 237).
149 Mbembe (2006, p. 156).
150 Mbembe (2019, p. 3).
151 Mbembe (2019, p. 40).
152 Mbembe (2019, pp. 13, 40). See also Tschakert (2022, p. 288).
153 Mbembe (2019, p. 27).

permanent severance of links with life and the living. . . . Death is a form of adventure and exploration of the infinitudes of life.'[154]

I began this chapter by observing that, in a house filled with small children, our world is populated by dragons, dinosaurs, unicorns, talking bears, and fairies. That is true, but I could equally observe that our house is filled with bugs, viruses, and bacteria. Since I've been reading more-than-human climate fiction, our house has hosted mice, moths, and carpet bugs, in addition to surprising numbers of daddy-long-legs. I had Covid-19 at least twice while researching this book. My children have brought home hand, foot, and mouth disease, tummy bugs, nits, viral infections of the ear, chest, and eye . . . more nits, chickenpox, threadworms, coughs, sneezes, warts, and ringworm. I am now closely acquainted with the life-cycle of the head louse, but this knowledge is instrumentalized in the grim purpose of expelling them from their homes in our hair, if only ever temporarily. We share our lives with myriad others, mostly unknowing but sometimes resentfully, inhospitably. The microorganisms—bacteria and other microbes—in our bodies contribute more genes responsible for human survival than humans do. Their cells outnumber human cells by ten to one—should I credit them as co-authors of my research?[155] Cudworth and Hobden suggest that 'if our task is to try and learn to live and die well together, in our troubled present on a damaged earth, then we need to consider what relations we might choose to make or retain and those we also choose to sever in trying to promote multispecies flourishing'.[156] It is impossible to live without doing harm. By our actions and our inactions violence is perpetuated and some relations are nurtured while others are truncated. Africanfuturist climate fiction can help defamiliarize these decisions and envision other ways to live convivially—if not pacifically—alongside others.

Conclusion

In this chapter I have argued that Africanfuturist climate fiction contributes to wider and deeper explorations of more-than-human futures than often takes place within dominant ecopolitical imaginaries. Fiction can be a powerful way of exploring alternative human relationships with nonhuman others, but fiction can also reinforce stereotypes and prejudices. Children's stories offer plentiful examples of this, from Enid Blyton's golliwogs to Roald Dahl's Oompa Loompas. Noël Sturgeon has provided an incisive critique of the gendered,

[154] Nyamnjoh (2017 p. 257). See also Plumwood (2004).
[155] Alberro (2024, p. 62).
[156] Cudworth and Hobden (2022, p. 666).

homophobic, and racialized representation of 'uncle Scar' in Disney's 1994 classic, *The Lion King*, surely one of the most globally influential portrayals of African landscapes and animals.[157] The 2019 remake of *The Lion King*, one of the most expensive—and commercially successful—films ever made, with a budget of around $260 million and grossing over $1.6 billion worldwide, portrays the heart-rending destruction of the lions' pride lands as a result of overconsumption and the unequal distribution of resources at the paws of the tyrant Scar.[158] It missed an opportunity, according to critic Greta Moran, to connect this ecological destruction to climate change but it does carry an optimistic message about the future. When Simba bemoans his inability to change the past, Timon, the optimistic meerkat, chimes in, '[b]ut you know what you can change? The future.'[159] We can also change the stories we tell, and how we represent and live alongside others.

Fictional stories are powerful ways of representing human relationships with nonhumans, offering imaginative possibilities largely unavailable in other formats. The Africanfuturist climate fiction stories examined here make a distinctive contribution to representing alternative human relationships with nonhuman others in three significant ways. First, through narrating ways of restoring better ethical relations with nonhumans, or good Land relations, they can strengthen and enrich attempts to explore 'interspecies diplomacy' within International Politics.[160] Second, through imagining posthuman futures in the form of shape-shifting and transformations into animals, these stories can destabilize the hierarchies through which diverse other subjects have been denied the status of full humanity—women, transgender, intersex, queer, disabled, neurodiverse, racialized, and others—and arrive at more capacious and inclusive imagined subjectivities for climate-changed futures. In so doing there is a 'refusal of purity politics' in these stories that resonates with queer environmental theory, which has drawn attention to the diversity of gender and sexual behaviours exhibited in the 'natural' world.[161] Third, by insisting on the presence of politics, violence, and power relations amidst future forms of more-than-human conviviality, these stories draw attention to the importance of exercising agency in how we live and die alongside others, albeit not in circumstances of our own choosing.

The political implications of these representations of nonhuman, posthuman, and more-than-human futures need further unpacking, and this is the

157 Sturgeon (2004, p. 268).
158 Allers and Minkoff (1994).
159 Moran (2019).
160 Constantinou and Christodoulou (2024); Fougner (2023); Opondo (2022); Strausz (2024).
161 Seymour (2018, p. 232). See also Butler (2017); Gaard (2017, p. 31); Seymour (2017).

task of the next chapter. Here, I want to make three concluding observations. First, the implication I draw from Africanfuturist stories about better ethical relations with nonhumans, such as in *Zoo City* or *Pumzi*, is that this will involve sacrifices in some form and attempts to make reparations for harms done in the past.[162] Second, an implication to be drawn from stories about posthuman shape-shifting as in *Who Fears Death* and *Lagoon* is that embracing human difference—racial, gendered, sexual, differently-abled, neuro-diverse—will be vital in coming to terms with changing climates. Third, the challenges of living and dying well alongside radical Others could mean a position of critical support to attempts to design political and legal institutions that respect the rights of rivers, forests, and deserts, as well as interrogating further what these rights mean in terms of the lives and deaths of less obviously charismatic species.[163]

It is easy to be an advocate of ecological activism when the nonhumans we have in mind are polar bears, butterflies, trout, panda, and rhino. It seems to be harder to advocate for the crocodile, vulture, head-louse, or bedbug.[164] I hope my son's love for the crafty croc will survive the gradual disenchantment of his world and that he will enjoying reading *Zoo City* one day. It is quite striking that the emblematic nonhumans in Africanfuturist climate fiction are spiders, swordfish, vultures, sloths, and crocodiles. The hostile, predatory, threatening representations of crocs in *Mara and Dann*, 'More Sea than Tar' and *Zoo City* provide a very different image of more-than-human futures than is usually imagined in climate fiction. Even here, however, there are glimpses of possible conviviality, albeit a conviviality that remains painful, violent, and dangerous. At the end of *Zoo City* the albino crocodile decides to assist Zinzi. 'It jerks its head at me. I flinch, but it's not attacking, it's motioning towards the stairs. For me to go.'[165] It then viciously attacks and kills the villain Odi Huron, providing a rare moment of justice in a story in which many of the other characters get away with their crimes. The final scene of the book is Zinzi leaving South Africa via the Beit border bridge to try to find Benoît's wife and children. She looks out at the river that marks the border, observing that '[t]here are crocodiles in that river'. But she crosses anyway, deciding that '[i]t's going to be the best thing I've done with my miserable life. And after that? Maybe I'll get lost for a while.'[166]

[162] Büscher and Fletcher (2020, p. 68); Sze (2020).
[163] Alberro (2024); Cudworth and Hobden (2022); Mitchell (2023); Tschakert (2022); Youatt (2014).
[164] Mitchell (2016); Tschakert (2022, p. 290).
[165] Beukes (2010, p. 349).
[166] Beukes (2010, pp. 355, 358).

7
Reparative Climate Politics and Reimagined Subjectivities

The contemporary politics of climate change are scary and violent. Weather patterns are already highly unsettled, causing misery to millions and predictions for the future are dire. For the most part, it appears that the strategies set out by states to mitigate and adapt to climate change are inadequate and unimaginative. Large-scale transformation of global fossil fuel economies, on the scale needed to mitigate the worst forecasts, are obstructed by some of the most powerful corporations, states, vested interests, and entrenched social practices that human history has produced. Given all this, it is unsurprising that many of these Africanfuturist stories narrate highly politicized ecopolitical imaginaries centring on revolution, revenge, reparation, and restoration. Wouldn't it be wonderful if we could find a switch that turns off violent and unsustainable power relations? It would be *fantastic*! In Nnedi Okorafor's *Noor*, the cybernetically enhanced heroine, AO, accomplishes something akin to this at the end of the story. The effects of multiple shocks to her neural implants somehow give her the ability to connect remotely and turn off the many massive desert wind turbines—the noor of the title—that are powering the Nigerian state and the malevolent Ultimate Corps. This brings peace to the desert by ending the Red Eye sandstorm. The bones that were flying endlessly in the wind 'fell to the ground. All those people could rest. And as the sand fell, all those people were also buried.' In the last line of the story we are told that 'in several big cities in a country far far away, the lights went out'.[1]

This feels like a moment of triumph, a victory for AO against the machines and the exploitative system of global hyper-capitalism that they power. This is akin to the blowing up of the pipelines that Andreas Malm has called for.[2] Read politically, however, it is also somewhat unsatisfying. The politics of climate justice require more than just turning off the power, even when it is violent and extractive. Keeping the oil in the soil and the coal in the hole is not

[1] Okorafor (2021, pp. 210–11).
[2] Malm (2021).

African Climate Futures. Carl Death, Oxford University Press. © Carl Death (2025).
DOI: 10.1093/9780198960775.003.0007

enough. More hospitable, egalitarian and sustainable societies and economies need to be imagined, repaired, and built. Nalo Hopkinson, when asked what she thought of Audre Lorde's comments about not using the 'massa's tools to dismantle the massa's house', responded that 'in fact, I don't want to destroy it so much as I want to undertake massive renovations [and] build me a house of my own'.[3] Reading all these stories about African climate futures reinforces the crucial point that perspective matters. Whose stories are told and heard is political. Hearing the stories of those living in the far away cities when the lights went out at the end of *Noor* would lead us to view AO's victory from a very different perspective.

This chapter asks: what are the political implications of these stories? What does 'victory' look like and who will take us there? What—or who—are the obstacles to be overcome? Recalling the policy visions and strategies discussed in Chapter 2 we can ask again, how are better climate futures imagined, envisioned and narrated? The politics of Africanfuturist climate fiction starts with a strike against the status quo, against those obstructing climate justice and social transformation. But we can go beyond that. Climate fiction can envision the political economies of the future, the societies and economies that come after the revolution. Fictional narratives also bring to life the political subjects and subjectivities of the future, the heroes to come and the humans and posthumans of climate-changed worlds. Questions of politics are 'ultimately questions of change, of desirable futures'.[4] While *Noor* ends at the moment the power is turned off, some of the stories examined in this book, including others by Okorafor, explore alternative socio-economic systems.

Politics has of course been at the heart of everything discussed in this book so far. The ways in which the places of the future are imagined, the relationships between past, present, and future, and the manner in which humans and nonhuman live alongside each other are all deeply political. The policy visions of low carbon transitions examined in Chapter 2 are also inescapably political. Yet, the texts of these policy imaginaries are often technical, institutionalized, and depoliticized.[5] They rarely ask, who are the winners and losers? Will the processes of change be democratic and how will they overcome entrenched vested interests? To what extent will low or zero carbon societies be post-growth or post-capitalist? Stories have a role to play in both imagining alternatives and in envisioning ways to fight against the status quo. As one of the most influential transitions scholars has argued, rather than constantly telling 'David versus Goliath' storylines 'in which heroic green innovations

[3] Hopkinson (2004, p. 8).
[4] Grosz (2004, p. 253).
[5] Feola (2020); Harris (2017, p. 179); MacGregor (2014); Paterson (2021a).

overthrow the giant', we need to hear more storylines that can show how Goliath can be 'weakened, eroded and destabilised, to enhance the chances of green Davids'.[6]

At the crux of the argument throughout this book has been the claim that Africanfuturist climate fiction can make a powerful contribution to telling these sorts of stories and envisioning the political struggles at the heart of climate futures. This chapter advances this argument in three parts. The first section shows how most of the stories examined in this book are radical and revolutionary in their politics, and as such they can contribute to learning 'how to fight all over again'.[7] This shared revolutionary outlook is read against Doris Lessing's *Mara and Dann* novels, where continuity and change are tightly bound together and the contingency and fallacy of utopian plans are foregrounded. This is a useful supplement to revolutionary Africanfuturism: another world is possible, but...

The second part of the chapter asks: what comes next? There are powerful reparatory themes in many of these novels, lending imaginative heft to calls for reparation ecologies, paying climate debts, and addressing loss and damage.[8] There are also a wide range of imagined alternative political economic structures. Most frequently, there are critical portrayals of hyper-modernist futures dominated by scientific and technologically advanced necropolitical capitalism. There are also several neo-feudal tales of warlords and chieftains. Some of these stories offer enticing glimpses of eco-socialist or post-capitalist political economies. In this respect they contribute to climate justice imaginaries and offer inspiring visions of what a reparative future could be.[9] In contrast to the more programmatic stories told by authors like Kim Stanley Robinson, however, I suggest that the greatest strength of Africanfuturist climate fiction lies in its potential to reimagine human subjectivity in future climate-changed contexts.

The third section of the chapter develops this argument. Many of these stories feature marginal, queer, outcast heroines, who respond to Fanon's call to 'introduce invention into existence'.[10] This section highlights the prominence of radical feminist, queer, and cyborg characters in stories like *It Doesn't Have to Be This Way*, 'Spider the Artist', *Noor*, and *Mara and Dann*.[11] These transgressive characters are some of the most powerful contributions of

[6] Geels (2014, p. 37). See also Albert (2024).
[7] Malm (2021, p. 22). See also Selbin (2019).
[8] Patel and Moore (2017); Sze (2020); Táíwò (2022).
[9] These groupings of ecopolitical imaginaries have close parallels to the more realist future pathways discussed by Michael Albert (2024), namely his collapse, techno-leviathan, and ecosocialist scenarios.
[10] Fanon (2008, p. 180).
[11] Lessing (1999, 2006); Mackay (2022); Okorafor (2011, 2021).

Africanfuturist climate fiction, but they must be read with one eye on the risk of over-individualizing how we imagine political change.[12] As Ghosh has suggested, '[j]ust as novels have come to be seen as narratives of identity, so too has politics become, for many, a search for personal authenticity, a journey of self-discovery', and he argues that Anthropocene politics demands much more imaginative posthuman renderings of agency.[13] There are risks associated with stories about messianic individuals who ignite revolutionary change, but overall I argue that rather than reaffirming liberal individualism these Africanfuturist stories more often work to challenge 'the meaning and significance of being (human)'.[14] These heroines stand for those who have so often been denied the status of full humanity, deemed to fall short of the standard of white, straight, 'Man'. In this chapter I show how these stories offer fantastic and futuristic ways of being Otherwise. Significantly, a repeated form of political agency in these tales is the power to write and rewrite stories themselves. This brings us back to some of the questions with which this book began: why read, write, and discuss speculative fiction at all?

Revolutionary Politics

The majority of these Africanfuturist climate stories have revolutionary visions at their heart. These are tales in which the world is turned upside down, Africa is at the centre of global politics, the future reshapes the past, nonhumans are political creatures, and humans are rendered posthuman. A revolutionary narrative needs an adversary to be overcome and in stark contrast to depoliticized policy visions, fictional stories are very good at narrating a 'struggle against an enemy' and the confrontations with adversaries that characterize contemporary necropolitics.[15] As Eric Selbin suggests, '[r]evolutionary imaginaries are the prerequisite for revolutions'.[16]

Nnedi Okorafor is the revolutionary tale-spinner par excellence. Radical change is at the heart of all her fiction. Her stories insist on the possibility and necessity of radical change and re-creation at all levels—ideologies, identities, social relations, forms of production, and consumption—much as Fanon argued that for true decolonization, 'everything must be started again

[12] MacGregor (2014); Maniates (2001).
[13] Ghosh (2016, p. 127).
[14] Jackson (2020, p. 1).
[15] Mbembe (2019, p. 42).
[16] Selbin (2019, p. 493).

from scratch.'[17] Onyesonwu, the central character in *Who Fears Death*, is a shape-shifter whose 'very essence was change and defiance.'[18] Her messianic role is to sweep away the existing order, to rewrite the Great Book and inaugurate a new past, present, and future. This manifests as a terrifying power that is violently destructive—'[a]ll things can burn, for all things must return to dust'—but it is also creative and emancipatory. As a result, '[t]his place will never be the same. Slavery here is over.'[19] The prequel to *Who Fears Death*, *The Book of Phoenix*, expands upon this revolutionary tale to explain how the twenty-first century American world order was brought down by Phoenix Okore—'beacon, and reaper, life and death, hope and redemption'—sent by the god Ani to raze everything in a terrible fire. 'Let me be the villain for the sake of justice', she proclaims. Her mentor tells her: 'you are change, Phoenix. Wherever you go, you bring revolution.'[20]

Lagoon is also a story of revolutionary change inaugurating a new world order, albeit one that invokes and remakes indigenous traditions, ghosts, and old gods. The alien arrival in Lagos heralds change and like Onyesonwu they are defined in terms of change itself. Ayodele tells Adaora directly: 'we are change', and as a result of their arrival '[t]he world had changed. Lagos was eating itself.'[21] One character, Fisayo, proclaims, '[t]his was the rapture, the apocalypse, the end.' A mysterious stranger—with an eerie resemblance to other old, white and sagacious characters in Okorafor's stories—wryly observes that '[i]t's going to get really interesting here soon, you'll see. It's a great time to be in Africa.' After Ayodele's sacrifice Adaora realizes that 'Lagos will never be the same.' Following his meeting with the alien elders, the Nigerian President tells the people: '[f]or the first time since we cast off the shackles of colonialism, over a half-century ago, since we rolled through decades of corruption and internal struggle, we have reached the tipping point. . . . We will be powerful again, o!'[22]

In these novels Okorafor draws upon the revolutionary, radical utopian tradition in Black political thought and Afrofuturism.[23] This tradition is also invoked in the *Black Panther* films, although their revolutionary utopianism is much more circumscribed and less radical than in Okorafor's novels. Achille

[17] Fanon (2001, p. 142); see also Agathangelou and Killian (2016). On Okorafor as a revolutionary fabulist, see Burnett (2015, p. 139) and Kotecki (2020).
[18] Okorafor (2018, p. 383).
[19] Okorafor (2018, pp. 369, 381).
[20] Okorafor (2015a, pp. 95, 105, 180)
[21] Okorafor (2014, pp. 39, 177).
[22] Okorafor (2014, pp. 127, 213, 270, 277–8). See also Hugo (2017, p. 48) and O'Connell (2016).
[23] Kelley (2002); Womack (2013); Zamalin (2019).

Mbembe pre-empted the core storyline of *Black Panther* when he described the myth of Africa as a land of 'unfathomable natural riches', which would 'emerge from its cave' to save the diaspora and humanity.[24] These films offer up the hope of a new world order, encapsulated in T'Challa's final speech to the United Nations at the end of the first film.[25] The Black Panther offers to share Wakanda's resources and knowledge with the rest of the international community: 'Wakanda will no long watch from the shadows. We cannot. We must not. We will work to be an example of how we as brothers and sisters on this earth should treat each other.... More connects us than separates us.'[26] He articulates a liberal vision of cosmopolitan tolerance and inclusion, channelling the spirit of Nelson Mandela and Martin Luther King in contrast to the firebrand revolution of Malcom X. A race-based politics of restitution and revenge is explicitly discredited in the film through the characters of N'Jobu and his son Killmonger.[27] The cosmopolitan vision of T'Challa is complicated in the sequel, *Black Panther: Wakanda Forever*, which foregrounds the limits of liberal inclusion and the broken promises of the West. Despite fighting each other for much of the film, eventually there is a truce between the Wakandans and the underwater Talokan kingdom, offering the potential for a postcolonial inter-continental allegiance between Africa and Latin America against the dominant world order.[28]

The short film *Pumzi* also imagines a transformation of the existing order. The central character, Asha, dreams visions of a greener future and the literal and metaphoric seed of renewal is carried out of the closed, homeostatic compound and into the wilderness, where it is sheltered and watered by Asha as she lies down next to the seedling.[29] The film's revolutionary message is emphatically confirmed by the final camera shot which pans out to reveal a vast forest that exists in what had been assumed to be a dead wasteland. Revolutionary political changes also take place in the short stories 'Bittersweet are the Waters' and 'The Drought', and there are revolutionary movements in the novels *It Doesn't Have to Be This Way* and *War Girls*.[30] These all reflect the prominence of revolutionary politics in the Africanfuturist genre.

Revolutionary change is dealt with somewhat differently in Doris Lessing's *Mara and Dann* novels. Lessing is not usually identified with the

[24] Mbembe (2017, p. 26).
[25] Coogler (2018).
[26] Coogler (2018).
[27] Iheka (2021, p. 222); Kotecki (2020, p. 165); Zeleza (2018).
[28] Coogler (2022).
[29] Hugo (2017); Omelsky (2014).
[30] Diene (2022); Mackay (2022); Onyebuchi (2019); Wilson (2019).

Africanfuturist tradition and while she was once an active member of the Communist party, she left and subsequently articulated a complicated blend of bleak realism and stubborn utopianism.[31] Her perspective is often interpreted as one who gave up on revolutionary politics and a political reading of her *Mara and Dann* novels seems to cast doubt on whether she thinks that things can ever really change. As a result, some critics have lamented the political implications of these novels, reading them as conservative, weakly ecological, and racist in places.[32] However, this is an overly simplistic reading of Lessing's politics and misses the fact that she remained firmly convinced that the most vital philosophical traditions were those like Marxism that attempted to look 'at things as a whole and in relation to each other', to imagine a 'world ethic' and insist upon addressing universal problems of discrimination and power.[33] Lessing is also a visionary of revolutions, although she was more concerned with the problems of 'a certain strand of revolutionary thinking, which is when you get a revolution or a change of government and your side gets in, then everything is going to be alright. Everything is not going to be alright.'[34]

In the *Mara and Dann* stories change and the mutability of life is a pervasive theme. Mara describes constant shifts in climate where 'everything always is a stage, one way of being changes into another'.[35] These changes apply to the human characters and civilizations of Ifrik as well, and much of the compelling force of the *Mara and Dann* novels comes from the charting of the slow, profound character development of the eponymous heroes in parallel to the societies they move through, many of which struggle to adapt to the pace of environmental change. One of the themes of the book is the difficulty of thinking about human political struggle on such a scale, and the meaning of change, history, memory, past, and future for continents and civilizations. As one character sadly observes, reflecting on the collapse of the great cities of Yerrup under the advancing ice, 'people always have a tendency to believe that what they have is going to continue for ever'.[36] Lessing seems to suggest in the *Mara and Dann* novels that change is the only constant, but it *is* a constant. '"There have always been changes in how the families are friends and enemies". "Always", whispered Mara, holding back her tears.'[37] It is the failure of certain groups—such as the Mahondi Kin in Chelops or the islanders in the

[31] Bertelsen (1986); Sargeant (2016); Watkins (2010).
[32] Becket (2009). See preceding discussion in Chapters 5 and 6.
[33] Walder (2008, p. 113).
[34] Quoted in Bertelsen (1986, p. 140).
[35] Lessing (1999, p. 322).
[36] Lessing (1999, p. 373).
[37] Lessing (1999, p. 35, see also pp. 306–7). See also Walder (2008, p. 104).

Middle Sea—to recognize and adapt to change that is one of the central political warnings in the book, a warning that is very relevant for contemporary climate change politics.[38] This failure to recognize change and adapt is also portrayed as a cyclical feature of human society, as the ancient civilizations of Yerrup failed to reform either their own technology-fuelled destruction or adjust to the coming of the ice: '[w]hile this process was going on they were hardly aware of what was happening, though a few did know and tried to warn the others.[39] Human civilization and the will to knowledge persists but Dann wonders if the effort is worth it: 'Dann, sir, what has been made can be made again", said Griot. "And again, and again, and again", said Dann.'[40]

Lessing allows her characters to imagine the possibility of a political revolution, a return of the Mahondi royal house to power, but it never seems likely. It is clear that there will be no 'happy-ever-after' or satisfying resolution.[41] The first novel, *Mara and Dann*, finishes on a note of fragile peace as the characters try to make a new family together on the Farm, after rejecting their heroic destiny and the role of Prince and Princess at the head of a new Mahondi empire. The fragility of this peace is emphasized, with tensions between Mara, Dann and Kira already breaking out into open conflict. The final lines are Dann asking Mara if her first thought every morning is: 'how far you're going to go today, one foot after another, another little bit of the way up Ifrik? And the two of us together?' Mara replies 'yes, it's true, but...'[42] This 'but...' hanging in the air at the end of 400 pages, could be taken as emblematic of the ambiguous, equivocal, contingent politics of the books. Change is possible, Lessing, seems to say, but... We can learn from the past, but... We must go on, but....

This reading is confirmed by the conclusion to the sequel, *The Story of General Dann and Mara's Daughter, Griot and the Snow Dog*. Dann and his companions leave The Centre and successfully invade neighbouring Tundra, building a new state dedicated to learning and wisdom.[43] Ali tells Tamar, Mara's daughter, to 'forget the long ago. We are here and we are now, and that is all. We are making a new start.' The final scenes appear to offer fulfilment and peace: '[t]ime had passed, measured by that tall girl walking down there in the garden.' Tamar is growing into maturity. Immediately, however, Lessing qualifies any sense of reassurance about the future with the news of another loss. Tamar rests 'her hand on the head of a white animal who was not Ruff. She

[38] Lessing (2006, p. 48).
[39] Lessing (1999, p. 381).
[40] Lessing (2006, p. 251; see also p. 192).
[41] Lessing (1999, pp. 340, 401). See also Pathak (2013, p. 2230); Watkins (2006, p. 255).
[42] Lessing (1999, p. 407). See also Watkins (2010, pp. 131–2).
[43] Lessing (2006, pp. 219, 272).

had wept so long over the loss of Ruff that a snow pup had been found, and he became her companion.'[44] Moreover, the constant sense of danger has not disappeared. Kira and her daughter, Rhea, are making threats against Dann's rule although Griot sees no real risk from them. There is a rumour that a kingdom to the south has plans to invade. Griot says that would be stupid. In the final lines of the book, Dann agrees: 'Well, yes, Griot, it would certainly be stupid. I agree with you there.'[45] With these words the epic saga ends by insisting on the stupidity of humankind and the suggestion that there is no escape from cycles of conflict or the loss of knowledge. Another world is possible, but....

Many of the stories examined in this book share this concern with revolutionary politics. Many insist upon the need to wipe the slate clean, start anew, and forge a new society. Many—like Lessing—examine the trials and tribulations of revolutionary politics too.[46] As famously observed in the film, *The Battle of Algiers*, '[i]t is difficult to start a revolution, more difficult to sustain it. But it's later, when we've won, that the real difficulties will begin.'[47] In this respect, these Africanfuturist climate fiction stories invoke a Black utopian tradition that emphasizes rupture and revolutionary change and the prospect of a new beginning.[48] Such visions of the future have various political consequences. One potential lesson for readers might be 'that violence and illegal activities may be both legitimate and necessary to pursue justice'.[49] Another lesson might be the importance of campaigns for repair, restoration, and reparations. The next section begins by considering this, and goes on to discuss the broader question: what comes after the revolution?

Climate Reparations and the Political Economies of the Future

I suggest that a central implication of these stories is to lend weight to calls for greater consideration of equity, justice, and historic responsibility in climate politics. They insist on African rights to development, control over land and resources, and reparations for slavery, imperialism, and climate coloniality.[50] In contrast to the—perhaps surprising—absence of these themes from

[44] Lessing (2006, pp. 267, 270).

[45] Lessing (2006, p. 282). See also Watkins (2006, pp. 258–9).

[46] Fanon (2001).

[47] Quoted in McClintock (1995, p. 388).

[48] Haiven (2014, p. 238); Iheka (2021, p. 27); Moonsamy (2016); Muñoz (2009); Samatar (2017); Sexton (2016); Weheliye (2014); Womack (2013, p. 41); Zamalin (2019).

[49] Jones and Parris (2018, p. 983). See also Malm (2021).

[50] Albert (2024, p. 167); Bassey (2012); Chigumadzi (2023); Jackson (2020, p. 214); Sabaratnam and Laffey (2023); Sultana (2022); Táíwò (2022).

African long-term climate strategies, these Africanfuturist stories are peppered with condemnations of imperialism, colonialism, and slavery, and conjure up ecopolitical imaginaries of revenge and reparation. In this sense they elaborate upon what Kelley refers to as the 'four-hundred-year-old dream of payback for slavery and Jim Crow', or what Raj Patel and Jason W. Moore (2017) call 'reparation ecologies'.[51]

Okorafor's terrible heroines—including AO, Phoenix, and Onyesonwu—all enact a violent retribution for the crimes committed against them and others, leaving '*bodies in our wake. Fields of bodies*', in order to 'wipe the slate clean'.[52] In Tsamaase's 'Eclipse our Sins' a vengeful Earth condemns humans for their crimes: 'carbon emission, racism, oil spills, sexism, deforestation, misogynism, xenophobia, murder.... Mother Earth no longer stands for such pollutants anymore.'[53] In Wilson's 'The Drought' a profound shift in human consciousness leads to 'the time for cleaning up their mess'.[54] Beukes' *Zoo City*, as discussed in the previous chapter, is centrally concerned with how characters pay their debts. Characters who refuse to live with the consequences of their deeds come violently undone in the end.[55] In many other stories the traumatic consequences of previous crimes resurface for individuals and societies and must be repaired.[56]

These stories resonate with calls from environmental and climate justice movements to prioritize reparations. Political theorist Olúfẹmi O. Táíwò sees climate reparations as a way to 'remake the world system'.[57] Patel and Moore argue that reparation ecology is a way of 'learning to interact with the web of life differently', and includes recognition, reparation, redistribution, reimagination, and recreation.[58] For Panashe Chigumadzi, 'Black reparations are impossible within the terms of this world. If slavery is the material and metaphysical womb of the modern world, reparations will require nothing less than the end of this world.'[59] More prosaic calls for reparatory measures focus on diverse policies and mechanisms, such as loss and damage funding, corporate accountability and liability by fossil fuel companies, land restitution, basic

[51] Kelley (2002, p. 7); Patel and Moore (2017).
[52] Okorafor (2018, p. 370; 2015a, p. 221). See also Hoydis (2017).
[53] Tsamaase (2019).
[54] Wilson (2019).
[55] Beukes (2010).
[56] Blomkamp (2009); Coogler (2018, 2022); Diene (2022); Embleton (2015); Kahiu (2009); Llansó (2015); Mackay (2022); Ngemntu (2022); Okungbowa (2019); Onwualu (2019, 2022); Onyebuchi (2019).
[57] Táíwò (2022, p. 1). See also Bassey (2012); Davis and Todd (2017, p. 774); Sultana (2022); Sze (2020).
[58] Patel and Moore (2017, p. 207).
[59] Chigumadzi (2023). A similar argument is made by Andrews (2024).

income grants, the Green Climate Fund, fossil fuel divestment, community control of resources and new forms of direct and community-based democratic deliberation, research and decision-making.[60] At COP28 in December 2023 rich countries pledged a total of about $700 million to the Loss and Damage Fund, a historic breakthrough in finally operationalizing demands from the Global South for compensation for harms already done by climate change, yet woefully short of the sums required.[61] Estimations of the annual financial need for compensating loss and damage are predicted to approach $600 billion by 2030.[62] Given the history of missed targets for financing climate mitigation and adaptation, there is little guarantee—and perhaps little hope—that the new loss and damage mechanisms will be meaningful, especially as climate strategies in the USA, EU, and UK typically ignore or explicitly reject obligations to redress ecological debts, and climate strategies in the Global South are often reticent about making these demands.[63] For Fadhel Kaboub, '[c]limate finance is a climate debt owed by the historic polluters of the Global North to Global South countries that are on the front lines of climate change. The Global North is in default and is refusing to pay its debt.'[64]

However, as Kelley has stressed, reparations are not simply about paying a financial bill to account for wrongs done but should be 'part of a broad strategy to radically transform society'.[65] Money and resources are important but not sufficient, even if they did come nearer to the vast sums needed. Kelley warns that 'without at least a rudimentary critique of the capitalist culture that consumes us, even reparations can have disastrous consequences'.[66] It is in contributing to this cultural critique, in envisioning a world-remaking—even world-ending—ecopolitical imaginary, that Africanfuturist climate fiction can play an important counter-hegemonic role.

The critical potential of science fiction, at least according to theorists like Jameson and Freedman, is most clearly apparent in its propensity for imaginative world-remaking, or envisaging alternative political, social, and economic structures.[67] This is what Suvin terms 'a mapping of possible alternatives'.[68] Ever since Thomas More's vision of a society without private property, in

[60] Albert (2024, p. 235); Bassey (2012); Büscher and Fletcher (2020, p. 68); Grasso (2024) Okereke (2024, p. 8); Pheko (2024); Sylla et al. (2024); Táíwò (2022, pp. 173–83).
[61] Shirley et al. (2023).
[62] Rizk (2023).
[63] Sabaratnam and Laffey (2023, p. 174). See also Andrews (2024).
[64] Kaboub (2023).
[65] Kelley (2002, p. 129).
[66] Kelley (2002, p. 133). See also Patel and Moore (2017, p. 210); Rao (2020, pp. 125–6).
[67] Freedman (2000); Jameson (2007).
[68] Suvin (2016, p. 24).

which 'though no man has anything, yet they are all rich', utopian fiction has sought to imagine new worlds.[69] Authors like Ursula Le Guin, Octavia E. Butler, and Paolo Bacigalupi have offered particularly vivid maps of post-capitalist political economies, but it is Kim Stanley Robinson—a former student of Jameson—who is probably the most prolific and sustained visionary in this respect.[70] The Mars trilogy tells the story of several revolutions, post-revolutionary constitutional conferences, and post-capitalist, post-colonial future socio-economic structures as Mars moves out of the political orbit of an Earth that is wracked by poverty, conflict, and climate change.[71] *New York 2140* offers a critique of hedge-fund capitalism and imagines an insurgent city organized around co-operatives that ultimately threatens to overturn the lives of the wealthy through mass insurrection.[72] *Ministry of the Future* attempts to map out how the parallel forces of radical eco-terrorist violence and insider political-economic reformism can produce a new system of global governance and socialized finance in which nature is properly valued and future generations are represented in political structures.[73] The *Science in the Capital* trilogy and the *Orange County* trilogy have also received extensive critical attention for their utopian blueprints.[74]

In contrast to Robinson's work, Africanfuturist climate fiction does not offer as many detailed examples of post-revolutionary political economies. Most of these stories focus more on the upheaval and overthrowing of the dominant system than the meticulous mapping of what comes after. This focus is politically significant: like More himself, most utopian writers spend at least as much time diagnosing the problems of the present as imagining future solutions. Indeed, as Ashcroft observes, 'without the problems the solutions would have no point'.[75] Yet they do offer a range of speculative ecopolitical imaginaries, including hyper-modernist, neo-feudal, and post-capitalist political-economic structures.

The most frequent ecopolitical imaginary takes the form of hyper-modernist, science, and technology-driven political economies. Often these are highly unequal and exploitative, future societies that are rapaciously state-capitalist and violent. Examples include Totlo Tsamaase's short stories;

[69] More (2016, p. 285).
[70] Cole (2022); Hatzisavvidou (2024); Johns-Putra (2019); Schneider-Mayerson (2019); Thaler (2022); Trexler (2015).
[71] Robinson (1992, 1993, 1996).
[72] Robinson (2017).
[73] Robinson (2020).
[74] See discussions in Johns-Putra (2016, 2019); Mehnert (2016); Nikoleris et al. (2019); Schneider-Mayerson (2019); Thaler (2022); Trexler (2015); Weik von Mossner (2017).
[75] Ashcroft (2017, p. 8).

'Spider the Artist' and *Noor* by Nnedi Okorafor; *War Girls*; 'Bittersweet are the Waters'; 'The Serpent's Handmaiden'; and *It Doesn't Have to Be This Way*.[76] In *Noor*, for example, the story opens and closes with violent confrontations that take place in markets, symbolizing the commodifying logic that AO eventually seeks to overturn.[77] The novel explicitly critiques the violence of colonial capitalism that has expropriated African communal land to construct vast wind and solar power stations, making possible a society of surveillance and electronic control. The villain of the story is the almost pantomime Ultimate Corps: 'something more than human, by Allah. It's the beast, a djinn. Fire and air, insubstantial, but very real. Human beings created it, but they will never control it.'[78] The story is distinctive in how it provides a detailed account of the emergence of renewable energy on a massive scale and the establishment of an Afrocentric global political economy of energy, via clashes over land grabs in the Sahara and the rights of communities to benefit from new technologies. Despite some good intentions, these new technologies work to reinforce capital accumulation, violence, and inequality. Nefarious alliances between state and capital also provide the main adversaries in other Okorafor fables, including Shell and Chevron in 'Spider the Artist' and LifeGen Technologies (also known as The Big Eye) in *The Book of Phoenix*.[79]

Tsamaase's three short stories all share an ecopolitical imaginary of a hyper-modernist, techno-authoritarian, rapaciously state-capitalist, and violent future world. In 'Virtual Snapshots' people live mainly in digital worlds where '[m]achines beep our lives across limbs of time. We sleep in dark home-cells, little bulbs lighting our prison, and sweep through the door in our avatar versions.'[80] Sunlight, water, and fresh air are rationed and insufficient credit means death. In 'Eclipse our Sins' respirators and cooling suits are needed to survive the hostile world created by the sins of the past, consisting of 'sharp glamorised buildings glooming above shacks and poverty', where '[f]resh air is as rare as a non-manufactured human being.'[81] Perhaps most hauntingly, in 'Eco-Humans' bodies themselves are commodities and the only job the narrator can find is at 'Better Us Tech, allowing experiments on the tampering of my biology, with little remuneration. It's the only way I can fuel my body, to keep

[76] Diene (2022); Mackay (2022); Ngemntu (2022); Okorafor (2011, 2021); Onyebuchi (2019); Tsamaase (2016, 2019, 2022).

[77] Mcbride (2023, p. 17).

[78] Okorafor (2021, p. 61).

[79] Okorafor (2011, 2015a).

[80] Tsamaase (2016).

[81] Tsamaase (2019).

it running.'[82] These are bleak and terrifying visions of hyper-commoditized futures.

The films *Pumzi*, *District 9* and *Neptune Frost* offer similar portrayals of highly technological, restrictive, and often violent societies, although they also contain forms of resistance and insurgency and the seeds of alternative social systems, most notably in the emancipatory hacker encampment of *Neptune Frost*.[83] The Maitu compound in *Pumzi* is hierarchical, tightly controlled, and is powered by manual energy production machines including treadmills and rowing machines. Individual water allocations are closely monitored and dreams are suppressed. The community is run by women, but it is authoritarian and coercive. As in the stories by Tsamaase, these films are versions of the 'If this goes on…' cautionary tale.[84]

A second grouping of future visions offer a version of a neo-feudal political economy, not pre-modern but hyper-modern worlds of '[u]rban militias, private armies, armies of regional lords, private security firms and state armies'.[85] In these post-collapse worlds technology has completely or partially stalled, markets are ruled by localized physical violence as much as by commodities and capital, and any significant state-based rule of law has fallen apart. Examples of these include *Who Fears Death*; the short stories 'Dune Song' and 'More Sea than Tar'; and Doris Lessing's *Mara and Dann* novels.[86] In these stories power is wielded by regional warlords or village strongmen, drug dealers, and ethnic chiefs. *Who Fears Death* offers glimpses of comfortable and peaceful rural lifestyles and there are parts of Lessing's Ifrik that are more settled and pleasant than others, but the wider context in all these stories is one of unrestricted exploitation, war, violence, hunger, and loss.

The final category of stories includes those that offer glimpses of attractive, eco-socialist, post-capitalist structures. The degree to which these ecopolitical imaginaries are explicitly post-capitalist varies considerably. At one end of the spectrum are stories in which there are potentially utopian or heterotopian enclaves within a wider capitalist system. In the *Black Panther* films, for example, Wakanda is a technologically advanced but ecologically sustainable society, ruled by enlightened royal chiefs and scientific/technological experts in close communion with traditional elders and the ancestors.[87] In both *Noor* and *Who Fears Death* there are complex egalitarian societies living within sand-storms in the desert. The narrative in *Noor* culminates when AO and

[82] Tsamaase (2022).
[83] Blomkamp (2009); Kahiu (2009); Williams and Uzeyman (2021).
[84] Thaler (2022, p. 222).
[85] Mbembe (2019, p. 84).
[86] Ize-Iyamu (2019); Lessing (1999, 2006); Okorafor (2018); Okungbowa (2019).
[87] Coogler (2018, 2022).

DNA reach the Hour Glass inside the Red Eye storm, a 'refugee city in the shadows ... full of the people who fall through or don't fit'.[88] It is a place of freedom, ingenuity, communal living, and relative safety: 'people lived right out in the open, not a house in sight. And there was so much open space, privacy wasn't an issue. ... [t]here were wide oriental rugs spread over the soft sand in the home areas. People walked about without shoes'.[89] In *Who Fears Death* the egalitarian and polyamorous Vah people live inside a sandstorm, taking refuge from the conflict raging outside.[90]

Okorafor's novel *Lagoon* concludes with a vision of a post-capitalist ecopolitical imaginary. It is an interesting exception to several of her other stories as there is no corporate villain. Lagos is chaotic, violent, and filled with greedy, ignorant, and deluded characters, but the adversaries are inept individual men—a husband, a preacher, a general—rather than the socio-economic system. The novel is utopian in the sense that it offers up the hope of a radically different world to come. We aren't told much about what sort of society will emerge as the novel ends at the point where this change is beginning. The vision of the Nigerian president is one of renewed institutions, empowered experts, alien technology, and 'fresh ideas that we can combine with our own'.[91] He articulates a vision of renewed relations with the land and waters, where crops and biodiversity would flourish and Nigeria would become a model for a new way of living.[92] It is certainly not explicit, although it is perhaps implied, that this will be a post-capitalist society.

We get further glimpses of desirable—if also flawed—future societies that have achieved a measure of ecological sustainability and appear to have post-capitalist elements in the short stories 'Mango Republic' and 'What the Dead Man Said'. 'Mango Republic' imagines Lagos as a futuristic eco-city in which technology and benign government seem to have successfully tackled poverty, disease, and hunger. This is a society based around the provision of social needs, especially in terms of assigned housing units: '[e]veryone got what she or he deserved, what he or she needed'.[93] The city is technologically advanced and powered by solar and biomass, but 'work still defined Lagosians. In fact, we did more work than ever before.... Everything had changed but much still remained the same'.[94] In 'What the Dead Man Said', New Biafra has been re-covered with forest lands containing bio-engineered plants and wildlife

[88] Okorafor (2021, pp. 130–4).
[89] Okorafor (2021, p. 136).
[90] Okorafor (2018).
[91] Okorafor (2014, pp. 277–8). See also Eatough (2017, p. 250).
[92] Okorafor (2014, p. 279).
[93] Agbedeh (2013, p. 183).
[94] Agbedeh (2013, pp. 182–3).

over nearly 80 per cent of the country, beneath a 'massive monorail system that connects all our cities to bypass the pristine forests below'.[95] Eko Atlantic is now a megacity, guns have been banned at the turn of the 2140s, and AI implants at the base of the skull monitor neural and physical activity to alleviate stress. In both these stories social relations appear more communal and egalitarian and poverty appears to have been eradicated by social planning and technological advances. As such they have at least elements of a post-capitalist ecopolitical imaginary.

Several of these stories also insist that a post-capitalist society will only be possible through some form of reparation and therapy to redress the traumas of the past. Chinelo Onwualu's 'Letters to my Mother' offers a positive vision of sustainable communal living, in which idyllic rural homesteads have survived the climate wars and live close to nature, provisioned by extensive communal forest gardens that are tended with shared manual labour and 'constant genetic tweaking'.[96] Social structures are explicitly non-hierarchical and the narrator, Obeche, recalls that 'in the old world, people were confined to a single occupation, forced to choose one thing to which they would dedicate all their time and energy. My people would shudder at the idea. In our Homestead, we all engaged in many different activities, according to our interest and capacity, to keep the community running.' Of all the stories discussed in this book, this vision of safe and egalitarian communities nurtured by the 'green embrace' of a vast forest garden is one of the most attractive. As discussed in Chapter 4, however, this harmonious vision is threatened by the re-emergence of past traumas, in the form of a diary discovered by the archivist Obeche, which requires collective acts of healing. Ultimately Obeche manages to take 'her pain and mine' and share it with the community and the forest. She reflects on the traumatized writer of the diary: '[l]ike the transformation from waste to compost her grief became a memory rich with life, a catalyst that would help me and my people grow stronger'.[97]

Another utopian vision is narrated in Jessica Wilson's 'The Drought', which describes how South Africa transformed from late modern capitalism to an ecologically advanced, post-capitalist society. The experience of living through Cape Town's Day Zero in 2018, we are told, led to the spontaneous emergence of new forms of generosity and connection, 'messages of love and solidarity' and a 'deep recognition of collective vulnerability'.[98] Developments in artificial

[95] Onwualu (2019).
[96] Onwualu (2022, p. 66).
[97] Onwualu (2022, pp. 65–6, 79).
[98] Wilson (2019, pp. 32–3).

intelligence mean that '[p]eople could communicate with each other at will by switching on or off certain parts of their brains', and the redundancy of communications technology precipitated the collapse of extractive mining and resource wars. As a result, 'all power relations fell away.... Or so the story goes.'[99] The final line suggests we should read this naively fabulous account with some scepticism, perhaps as illustrative of how complex processes of social change tend to be narrated in highly simplified, almost magical ways, laying the foundations for traumatic re-eruptions of political and emotional conflict in the future.

This section has explored a range of visions of future political economies, spanning positive and negative hyper-modernist, science and technology-driven imaginaries, neo-feudalist post-collapse societies, and brief glimpses of potentially post-capitalist social systems. Rather than providing detailed 'solutions' these stories should be read as ways of defamiliarizing and estranging us from our current world; as Jameson explains, the core of utopian speculation is 'not the commitment to a specific machinery or blueprint, but, rather the commitment to imagining possible Utopias as such, in their greatest variety of forms'.[100] The next section argues that the main political contribution of Africanfuturist climate fiction consists in how it reimagines human and posthuman subjectivity in climate-changed futures.

Africanfuturist Eco-hero(in)es and Queer Subjectivity

One of the greatest strengths of fictional writing is imagining compelling characters in rich detail. Through narrative interrogation of identity and agency, fiction holds out the prospect of defamiliarizing dominant understandings of subjectivity.[101] This is a major contribution of fictional ecopolitical imaginaries, especially given the imaginative poverty of official long-term national climate strategies in terms of envisioning futures where actual people might live. The Africanfuturist climate fiction explored in this book makes three distinctive contributions to imagining new eco-heroines and climate-changed subjects. First, as discussed in Chapter 6, 'Dangerous ecologies and more-than-human conviviality', these stories help to reimagine the ethics and politics of human and nonhuman, more-than-human, and posthuman co-dependencies

[99] Wilson (2019, p. 33).

[100] Jameson (2007, p. 217). See Albert (2024) for an insightful discussion of the prospects for various socio-economic trajectories including a mapping of plausible ecosocialist 'concrete utopias'.

[101] Ghosh (2016); Milkoreit (2016); Nikoleris et al. (2019, p. 80).

and encounters. Second, by exploring feminist, queer, and cyborg perspectives on the forms of political subjectivity narrated in these stories, this section shows how destabilizing dominant binaries can allow us to imagine more diverse heroines for climate-changed futures. Third, rather than reinscribe individualist narratives of exceptional heroes, these stories suggest that rewriting pasts, presents, and futures is a form of political agency that is important but under-emphasized in dominant ecopolitical imaginaries.

There is much of interest for feminist, queer, and cyborg perspectives in how heroic agency is imagined in these stories. For example, at least two-thirds of the stories discussed in this book have lead characters that are strong women and challenge traditional gender roles and identities. These characters work to subvert the often patriarchal, masculine, and misogynist cultures of climate science, elite politics, and social movement activism.[102] They also challenge the ways in which women are presented in many works of science fiction and climate fiction. Very few are represented in 'traditional roles as nurturers and caretakers', or as the iconic 1960s and 1970s 'gun-slinging, baby-toting, Afro-coifed Amazon warrior'.[103]

Second, many of these stories feature queer characters, movements, and themes.[104] Mackay, for example, describes *It Doesn't Have to Be This Way* as 'the story of three queer friends trying to navigate an increasingly fractured, violent and unstable world'.[105] There are same-sex relationships, LGBTQ+ organizations and characters who self-identify as gay in many of the stories, for example the characters Jacobs, Seven, and Rome in *Lagoon*; Dann in *Mara and Dann*; Onyii and Chinelo in *War Girls*; and Viwe, Malcolm, and Luthando in *It Doesn't Have to Be This Way*.[106] Contrary to stereotypes of 'African culture' as homophobic and myths that non-normative gender and sexual identities are 'un-African', the prominence of queer themes in these stories demonstrates the power of fiction to question and oppose 'essentialist patriarchal hetrosexist heteronormative binary configurations of sexual orientations and

[102] This approach has been influenced by feminist theorists and researchers including Laduke (2004); Luxton (2018); MacGregor (2014); Nixon (2011); Sandilands (2017); Sturgeon (2004). For debates over the applicability and implications of the label 'feminist' in African contexts, see Fakier and Cock (2018); Krishnan (2014); Newell (1997); Nfah-Abbenyi (1997); Nyanzi (2014); Oku (2021); Tamale (2020); Wilson-Tagoe (1997).

[103] Kelley (2002, p. 136).

[104] Like the term 'feminist', there have been extensive debates about the applicability and implications of the term 'queer' in African contexts. See Bennett (2018); Brooks et al. (2022); Gore (2018); Green-Simms (2016); Hoad (2007); Matebeni and Pereira (2014); Nyanzi (2014); Olaoluwa (2018); Otu (2021); Rao (2020); Reddy et al. (2018).

[105] Mackay (2022).

[106] Lessing (1999, 2006); Mackay (2022); Okorafor (2014); Onyebuchi (2019). There are also the characters of Neptune and Matalusa in *Neptune Frost*, see Williams and Uzeyman (2021).

gender identities'.[107] For Cameron Butler, '[t]o queer is to question the categories, definitions, divisions, distinctions, and dualities that exist and that usually go unchallenged within society'.[108]

Third, there are also many prominent cyborg characters and themes, where humans are augmented with technology in transformative and transgressive ways. The figure of the cyborg represents, for Sandilands, the ways in which none of our bodies are really 'integral nor autonomous' but are interpenetrated with countless other (human and nonhuman) bodies, technologies and power relations.[109] Examples of stories with prominent cyborg themes include 'Spider the Artist'; *Noor*; *Black Panther*; *Pumzi*; *It Doesn't Have to Be This Way*; 'The Serpent's Handmaiden'; 'What the Dead Man Said'; *War Girls*; 'Virtual Snapshots'; 'Eco-humans' and 'The Drought'.[110] The genres of Afrofuturism and Africanfuturism have always been closely concerned with the relationship of technology to subjectivity, and Mbembe has argued that the manner in which 'the technological devices that saturate our lives have become extensions of ourselves' is itself prefigured by much older African traditions and philosophies of animism.[111] Africanfuturist climate stories draw upon these histories and politics to imagine characters who perform their sexuality, gender fluidity, (dis)ableism, and relationship with nonhumans in ways that subvert dominant norms about identity and agency.[112]

The most obviously queer story is *It Doesn't Have to Be This Way*, which begins with a gently satirical romantic plotline revolving around the intertwining love-lives and friendships of three South African gay men: Luthando, Viwe, and Malcolm. Luthando and Malcolm attend a middle-class 'greenie' tree-planting festival where they meet Viwe, and eventually Luthando and Viwe fall in love. Luthando's affinity for trees and forests develops into climate justice activism, organizing a protest march and ultimately participating in a direct action attack on Eskom's Medupi, '[o]ne of the world's largest and dirtiest coal-fired power plants'.[113] After the action is foiled, he is incarcerated and smeared by corporate and right-wing social media. Traumatized by his

[107] Nyanzi (2014, p. 61). See also Matebeni and Pereira (2014, p. 7) and Rao (2020, p. 16).
[108] Butler (2017, p. 275).
[109] Sandilands (2017, p. 234). See also Haraway (2016); Kafer (2013); McBride (2023); Melzer (2006); Moonsamy (2020).
[110] Coogler (2018, 2022); Kahiu (2009); Mackay (2022); Ngemntu (2022); Okorafor (2011, 2021); Onwualu (2019); Onyebuchi (2019); Tsamaase (2016, 2022); Wilson (2019). See also *District 9* and *Neptune Frost*.
[111] Mbembe (2019, p. 108). See also Eshun (2003); Samatar (2017); Womack (2013).
[112] Brooks et al. (2022); Fiskio (2012); Hall (2019); Haraway (2016); Jackson (2020); MacGregor (2014); Mcbride (2023); Oku (2021); Omelsky (2014); Otu (2021); Sandilands (2017).
[113] Mackay (2022).

experience of prison he withdraws into depression on his release, before his horrific betrayal by Viwe to homophobic vigilantes. Yet he remains the novel's voice of sanity, compassion, and humanity, raising his adopted son Milo and rescuing Malcolm. In a heated exchange with Viwe over self-worth and queer pride, Luthando advocates, laced with sarcasm, for something like a queer eco-sensibility: '[y]ou know, gays are probably the best thing to ever happen to this planet. We reduce overpopulation. We save the world from the toxic masculinity that got us into this mess—domination and conquest instead of nurturing, feminine qualities. We're nature's attempt to self-correct.'[114] While such a remark could be read as implying a problematic association between same-sex relationships and population control, in the context of the novel it can be seen as one snapshot of an ongoing struggle to reconcile conflicting identities and articulate a form of queer self-worth and mutual care in the context of an overtly hostile and violent society. It also resonates with the claim by Nicole Seymour that 'queer values—caring not (just) about the individual, the family, or one's descendants, but about the Other species and persons to whom one has no immediate relations—may be the most effective ecological values.'[115]

Luthando's capacity to persist and to keep caring for others carries the story to its conclusion. Luthando's letters from outside the Citadel make Malcolm—who is on the inside, living a life of apparent safety but increasing anxiety—realize there are still struggles worth fighting for. Rather than give up on human relationships in a world of climate enclaves and technological atomization, Malcolm decides to leave the Citadel and seek out Luthando. Tragically, he arrives too late. There is no happy reconciliation, and instead the book ends with Luthando's adopted son, Milo, leaving Malcolm and the temporary safety of his commune to search for his parents. As the heat rises, Milo finds sanctuary in a cave on Table Mountain, full of Luthando's paintings and stories.[116] Milo lies on the floor, trying to stay cool, and the final lines of the book recall Luthando's early obsession with plants, leaves, roots, and tendrils:

> My body tingles and sprouts. Little green tendrils come alive in the heat and extend from my fingertips, my toes, the backs of my knees. Plants are the thing to be. They don't envy; they don't doubt; they don't worry about their parents. I can hear them, the way they talk to one another. I grow long vines from every bone. Leaves spread over my skin, basking in the sunlight. Roots push down

[114] Mackay (2022). For elaboration on queer eco-sensibilities, see Azzarello (2016); Butler (2017); Evans (2017).

[115] Seymour (2013, p. 27). See also Mcbride (2023).

[116] Wynter and McKittrick (2015, p. 66).

> into the earth. They grow deeper and deeper, twisting and forking, anchoring me to the rock, holding me steady. I can't move. My body turns to soil. My skin glows green, photosynthesising. Sunlight courses through my veins. I am whole, I think. I am everything.[117]

This final note of transcendental, earthy, grounded hope resonates with the argument for a hopeful, queer utopianism as advocated by José Esteban Muñoz. For Muñoz, queerness is 'a desire for another way of being in both the world and time, a desire that resists mandates to accept that which is not enough.'[118]

Many other examples of troubled, transgressive, queer heroines can be found in Nnedi Okorafor's stories. The character AO, in *Noor*, has been disabled since an awful car accident as a child and has been repaired and reconstructed through augmentations including a cybernetic left arm, artificial limbs, and neural implants. She tells us that, 'by definition, to many Nigerians, I was trouble. Even in Abuja, though it wasn't as rabid as in the south, I was a demon. A witch. An abomination ... if you were one of those people who *seemed* to be "*more* machine than human" for whatever reason, one of those who "refused to obey the laws of nature and die", you were a demon.'[119] Her joints are hyper-flexible, her limbs can extend at will and are incredibly strong, and 'to save my parents the cost, I'd opted out of the "humanising exterior" that felt, smelled, even reacted like human flesh.' As a result, 'I am part machine. I am proud to be part machine. I was born twisted and strange by their standards. And after so much recovery, I was somehow amazing.' She grows to realize she has additional strange, inexplicable powers, abilities that seem to emerge alongside disability and pain: 'sometimes when something breaks, things come out of the cracks'. This gives her the power to broadcast the truth about Ultimate Corps to the Nigerian people in a cyberpunk snarl: 'Do I make you feeeeel uncomfortable? Unsafe?... You want to know why you feel that way? You want to know who's to blame for *me*?? The same people who made your mobile phones. The same people who made your children's toys. The company who sells you the super-meats you put in your husband's okra soup and pounded yam.'[120] Her uncanny, disruptive abilities enable her to shut down the noor wind turbines, ending the Red Eye sandstorm and derailing global capitalism.

[117] Mackay (2022).
[118] Muñoz (2009, p. 96).
[119] Okorafor (2021, p. 10).
[120] Okorafor (2021, pp. 15, 19, 121, 193).

Another deliciously, terribly counter-cultural story from Okorafor is 'Spider the Artist'. Eme, a young women living in the Niger Delta, has 'tried to give my husband some happiness' but they have not managed to have children and he is beating her.[121] At the end of the story she has found a happier relationship and is pregnant, but the twist is that the loving relationship is literally cyborg. The bond between Eme and the zombie droid, Udide, appears to bring both pleasure: 'I loved playing for it. It would listen. Its lovely eyes would glow with joy. Could a robot feel joy? I believed intelligent ones like this could.' Eme has misgivings because she knows that the zombie droids are violent creations built to guard oil pipelines in the Niger Delta. 'I'd befriended this artificial creature. I knew it. Or I thought I knew it. But what did I really know about why it did what it did? Or why it came to me?' Then Udide spins a musical instrument out from inside itself and plays for her. In the midst of a descent into violent war between humans and zombies, Eme 'dreamt about being embraced by deep blue liquid metal, webs of wire and music.... *Ah-ah*, this machine and me. You can't imagine.' At the end of the story, Eme is pregnant, and 'Udide tells me it is a girl. How can a robot know this? Udide and I play for her every day. I can only imagine how content she is. But what kind of world will I be bringing her into? Where only her mother and Udide stand between a flat out war between the Zombies and the human beings who created them?'[122] This is an overtly cyborg story of relationality and interpenetration, that conveys both the tenderness and the pain of our technologized, hybrid contemporary subjectivity, in which humans depend on machines—and other nonhumans—for our subsistence, art, pleasure, conflict, and reproduction.[123] From the perspective of queer theory, 'Spider the Artist' suggests that new sorts of loving families might offer the potential for a future in which beauty and creativity can exist, outside the violent confines of heteronormative patriarchy and anthropocentrism.[124]

We find another troubled, queer, anti-hero in Lessing's *Mara and Dann* novels. While Mara is forced to dress as a boy and pretend to be 'Maro' to avoid harm, it is Dann who is deeply traumatized by exile and violence from a young age, and these wounds are exacerbated by torture and sexual exploitation at the hands of other men.[125] He grows up into a successful and charismatic soldier and leader, inspiring devotion in his men and having sexual relationships with some of them. Although he has a child with Kira he doesn't love her and feels

[121] Okorafor (2011).
[122] Okorafor (2011).
[123] Haraway (2016); Sandilands (2017).
[124] Gaard (2004); Muñoz (2009); Olaoluwa (2018); Otu (2021); Seymour (2013).
[125] Lessing (1999). See also Kuznetski (2021).

trapped in the new extended family they establish. He leaves the Farm before his child is born and in the sequel—*The Story of General Dann and Mara's Daughter, Griot and the Snow Dog*—the story focuses on Dann and his desire for oblivion. The opening line sets the tone: 'A slight move to one side or the other—a mere hand's breadth—and Dann must fall.'[126] His repeated drive to seek out danger and death—including an addiction to opiates and a desire to get dangerously close to the calving, collapsing ice cliffs of Yerrup—can be read not only as trauma but also, potentially, as a manifestation of a 'queer environmentality', or a flaunting of the supposedly primary natural laws of survival and reproduction.[127] Provocatively, Lee Edelman has framed this as the queer 'death drive', a form of resistance to heteronormative 'reproductive futurism'.[128]

Whatever role his sexual desires play in his characterization, Dann is an unsettling and disturbing eco-hero. He has most purpose and energy when trying to persuade the communities living in the Middle Sea about the melting of the ice and the changing climate, but at other points in the stories he sells his sister into prostitution, abandons his child, and languishes in self-pity and despair.[129] Yet, eventually, such hope as Lessing permits us comes through the man and leader Dann has become. Dann is capable of tenderness, care, and love: for his sister, Mara; for the dog he saves, Ruff; for a Thores woman, Kass; and for his niece, Mara's daughter Tamar. He feels grief, loss, loneliness, and responsibility to such an extent that they frequently overwhelm him. Yet he keeps going.

Dann also has the capacity to inspire love in others. Griot is moved to tears when he watches Dann, Tamar, and Ruff on a stage addressing the soldiers. 'He was weeping: he had to, those three, Dann, so handsome, and the lovely child, so like him, and the snow dog, the shining animal with its white ruff into which Tamar's little hand had disappeared. A gleam of weak sunlight fell at this propitious moment from a wet cloud, and at first Dann, Tamar and the snow dog, and then the whole square, were touched with a gold light.'[130] When Griot asks himself why the soldiers love Dann, he wonders if they see 'something fine and strong. Dredged from long ago memories, perhaps not even his, or from old tales, a word came floating—noble. There was something noble in that poor sick creature out there. And what did it mean, saying that? Griot

[126] Lessing (2006, p. 1).
[127] Azzarello (2016, p. 4); Lessing (2006, p. 65).
[128] Edelman (2004). For responses to Edelman, many of which reject his arguments about the queer death drive, see Butler (2017); Evans (2017); Johns-Putra (2019, p. 22); Muñoz (2009); Rao (2014); Seymour (2013).
[129] Lessing (2006, p. 54).
[130] Lessing (2006, p. 219).

didn't know. He felt, in spite of everything, that he would die for Dann.'[131] His faith is ultimately proved right and the new kingdom of Tundra is peaceful and prosperous under Dann's rule. Yet Dann is still a complex hero: 'A tall, but stooping, smiling man ... his smile was something that Griot had to watch and wonder over. Ironic, yes, always; tender, kind ... but was there something else, too, a touch of ... was that cruelty? No, it was a carefully hidden and held-down impatience.'[132] As I will argue below, the great man and noble leader trope here is politically problematic, but even so I read the character of Dann, ultimately, as embodying what Muñoz describes as a faltering, tender, sensual, queer hope.[133] In contrast to those who have argued that the Mara and Dann novels are not 'eco-political', I suggest that through disturbing (anti)-heroes like Dann they offer a glimpse of the sorts of political agency, the sorts of hope, that can emerge from within the midst of pain, trauma, and disappointment.[134] For Julia Kuznetski, these stories work to undermine the 'paradigm of the sane, autonomous, and singular white hero'.[135] As such, Lessing's novels are a powerful contribution to '[p]ost-apocalyptic political thought' in a 'feminist vein' that helps imagine 'new political futures after man-made disasters'.[136]

These queer and/or cyborg eco-hero(in)es challenge our conceptions of revolutionary agency, pluralizing the sorts of subjects that could help to change our political, social, and economic structures. They embody a transformative politics that can substantially enrich and diversify dominant political imaginaries, and that questions the 'categories, definitions, divisions, distinctions, and dualities' of mainstream climate imaginaries.[137] Binaries such as human/machine, human/nonhuman, freedom/domination, past/future, hero/villain, good/evil, normal/deviant, and reason/madness are all challenged and rewritten in these stories. This is a significant contribution to climate fiction, given, as Johns-Putra has suggested, mainstream climate change novels tend to employ quite conventional literary strategies of world-building and character development, and often centre white, cis, straight men (as in stories by McEwen, McCarthy, Bacigalupi, Robinson, and many others).[138] Moreover, these stories can challenge conventional narratives about families and parenthood and their role in climate politics, imagining alternative family

[131] Lessing (2006, p. 247).
[132] Lessing (2006, p. 273).
[133] Muñoz (2009).
[134] Becket (2009, p. 141).
[135] Kuznetski (2021, p. 60).
[136] Hunt (2023, p. 837).
[137] Butler (2017, p. 275).
[138] Johns-Putra (2019, p. 37).

structures and subverting heteronormative reproductive futurism in interesting ways.[139] They foreground non-biological caring relationships that are tender, wounded, and infused with both hope and loss.

It is also worth reflecting, at this point, on the way that many of these Africanfuturist stories remain faithful to the traditional fictional reliance on a central heroic protagonist. Such fables tend to reproduce individualist accounts of change, and some of these stories—such as Okorafor's *The Book of Phoenix* and *Who Fears Death*, Ngemntu's 'The Serpent's Handmaiden' and Kahiu's *Pumzi*—are overtly messianic.[140] Griot's observation that Dann has an ancient nobility is significant here. Accounts of democratic struggles and collective organizing for change are rare in these stories, highlighting a wider difficulty in narrating collective politics.[141] In the KwaZulu-Natal of 2038 in 'The Serpent's Handmaiden', for example, we are told that 'the people were too meek to revolt' and the story culminates in deliverance from drought after Shwelezile—whose name is described as meaning she 'who brokers divine mercy' in Zulu—has a transcendental vision under an olive tree on a mountain.[142] This is problematic for critics like Ghosh, who has argued that we need 'to find a way out of the individualising imaginary in which we are trapped'.[143]

There are numerous examples of conservative forms of political agency or anti-democratic fables in these stories. Oligarchic elites drive history in *Black Panther* and *Mara and Dann*.[144] While the blockbuster films offer little to reassure progressive activists or democratic theorists, the representation of elite politics in Lessing's novels is complicated. On the surface these novels seem to offer a familiar hero-quest narrative in which royal children endure exile and tribulations before eventually fulfilling their historic destiny. Sargeant accuses Lessing of being 'unable to imagine the kind of communal action which is the necessary hinge between individual work and its envisaged global result'.[145] But Lessing articulates a deeper scepticism towards all forms of institutionalized power. When Mara asks General Daulis whether the Mahondi clan, who used to rule all Ifrik 300 years ago, ruled well, he laughs and answers, '[w]e ruled well from the point of view of the Mahondis'.[146] When Mara and her brother eventually arrive at The Centre and are pressed to fulfil their destiny as Princess

[139] Death (2024a).
[140] Kahiu (2009); Ngemntu (2022); Okorafor (2015a, 2018).
[141] Branch and Mampilly (2015); Haiven and Khasnabish (2014); Selbin (2019).
[142] Ngemntu (2022).
[143] Ghosh (2016, p. 135).
[144] Coogler (2018, 2022); Lessing (1999, 2006).
[145] Sargeant (2016, p. 120).
[146] Lessing (1999, p. 344).

Shahana and Prince Shahmand, the last of the Mahondi royal house, it is clear that this is a dying fantasy of the deluded inhabitants of The Centre. Mara reflects 'it's the same, wherever you see it, the Hadrons, or the Hennes, and—did she remember something of the sort in her own family, from her early childhood? Power. The ruthlessness, just hidden by smiles and courtesies.'[147] Dann openly questions why people treat him any differently to the thousands of other migrants and refugees around them. When Griot sings Dann's praises, Dann replies: 'You forget Mara and I were among so many—so many, Griot. Who praises them? Who sings about them?'[148] I read these stories as a critique of elite politics and the inherent human tendency to abuse positions of power, but neither do they offer much hope for radical democratic politics.

There are only a handful of sustained attempts to represent collective political action in these Africanfuturist stories. Okorafor's *Lagoon* is one example, where collective agency manifests through the many different perspectives—human and nonhuman—of the inhabitants of Lagos, weaving them together into a hymn for the 'heads and hearts of the people of Lagos.'[149] Yet, ultimately this energy for change flows through elitist institutions: the alien Elders negotiate directly, behind closed doors, with the Nigerian President, and radical social transformation spreads downwards from them, assisted by the expert advisors of Ayodele, Adaora, Anthony, and Agu, 'the trained officials he was appointing to take the lead in in keeping everyone safe.' While there is a degree of satirical scepticism towards Nigerian politics—the unnamed President is a 'shrivelled, sickly man'; the vice-president, Wishwell Williams, is roundly mocked; and other figures of authority are portrayed as greedy, violent, selfish, and ignorant—ultimately the forces of revolutionary change are channelled through the institutions of the state. There are subversive messages throughout—a giant underwater spider tells Adaora, Agu, and Anthony that 'even in the corners of palaces, spiders dwell ... remember that, if you ever find yourself walking in the halls of the great and powerful'—but the climax of the plot is the speech that the President makes to the nation after meeting the alien elders. He knows his time has come: '[e]veryone needed him to do this right. Everyone in Lagos. Everyone in Nigeria. Maybe everyone in the world. He worked best when people needed him.... Now, for the first time, he felt like a president. And this speech would be his first real act as Nigeria's *true* leader. Oh, it was exciting.'[150] In the wider context of Okorafor's work, there is a puzzling trust in discredited political elites at the end of *Lagoon*.

[147] Lessing (1999, p. 370).
[148] Lessing (2006, p. 275).
[149] Okorafor (2014, p. 194).
[150] Okorafor (2014, pp. 278, 83, 217, 233, 259, 275, respectively). See also Eatough (2017).

There are three stories worth examining that do pay closer attention to collective action and social struggle. First, Mackay's *It Doesn't Have to Be This Way* relates Luthando's radicalization from a tree-planting festival to guerilla gardening, then to a viral QR-code campaign and organizing a march through Cape Town, eventually leading a direct-action attack against a coal-fired power station. While none of these actions is particularly successful, this escalation of activism is mirrored elsewhere in the world. We are told that 'vigilante killings of climate change deniers ... have been taking place across the world. A retired petroleum executive in Texas was pulled out his house and burned alive.'[151] After the apocalyptic 'change', however, the dominant movements are violent, millenarian, religious fundamentalists such as the Shepherds, founded by Viwe. Collective action is certainly not romanticized in this cautionary tale, but the importance of acting collectively for more just climate futures is a key message. Milo recalls how Luthando has told him that 'too many people expect someone else to save them—a prophet, a president, a hero—but that's why the world is in the state that it's in. It's always someone else's problem to fix it. Someone else will get us out of this mess. We have to be the someone else, he says. We have to save ourselves.'[152]

More successful collective actions are portrayed in two other stories. First, in *The Book of Phoenix*, the SpeciMen are inspired by Phoenix Okore to break free from the towers. They wave Ghanaian flags, hack CNN to broadcast their revolt and call themselves the 'Ledussee' or 'Let us see'. They even manage to destroy one of the towers, leaving behind a note referencing Nat Turner and proclaiming that '*A luta continua*'.[153] While Phoenix herself is a messianic demi-god—as envisioned in Figure 6—she inspires collective action. Second, the short story 'Bittersweet are the Waters' features a full-blown anti-corporate revolution from below. Set in the sprawling new metropolis of Amanzi, on the old ocean floors of the Atlantic basin, we follow Miyongi as he assists a mysterious women-only water-cult called the Nymphomaniacs in an urban insurrection against 'H2OCorp'. Myongi harnesses the botanical wisdom of his grandmother and releases enhanced seeds that turn the grey city green with monstrous cacti. The city's inhabitants rise up, armed by the Nymphomaniacs, in a rebellion against H2OCorp. Brandishing machine guns and taking guerrilla gardening to fantastical heights, they erect barricades around their city: '[t]he world knew the neighborhood as Garden now, and it was impregnable'.[154] The revolution is successful and the cult is vindicated as rain arrives

151 Mackay (2022).
152 Mackay (2022).
153 Okorafor (2015a, pp. 155, 191–2).
154 Diene (2022, pp. 98–9).

Figure 6 Eric Battle, 'Seed'

for the first time in a millennium. But there is a cost: '[t]ens of thousands had died when his fairy-tale cacti sprouted.... They had fought hard, and millions had drowned.'[155]

These three examples bring to light a wider absence of convincing collective action in climate fiction more generally. The novel as a format—particularly as it emerged in its classic form in the nineteenth century—is closely tied to individual life-stories and character arcs, and it is hard to narrate collective action without losing the thread of the plot.[156] As an isolated example, Kim Stanley Robinson has gone furthest in portraying radical, popular climate-fuelled insurrection. He insists that 'there needs to be mass action' in order to realize sustainable, post-capitalist futures, and disdains stories and ideologies that over-emphasize 'the antics of individual humans'.[157]

In most of the stories discussed in this book, individual heroines—diverse, queer, cyborg, and posthuman—drive the plot and inaugurate change. In a political context where many of our leaders and organic intellectuals are struggling to inspire much hope, perhaps the desire for new heroic actors is understandable. However, it is also worth emphasizing that these stories do not really invoke a liberal, modernist faith in autonomous individual leaders, but rather they locate their counter-cultural heroines within webs of planetary entanglements.[158] This is especially evident in the more-than-human transformations wrought by Onyesonwu in *Who Fears Death* and Phoenix in *The Book of Phoenix*.[159] Moreover, it is worth paying close attention to the ways in which heroic agency manifests in these stories. Many of these central characters deliver change through the act of writing and rewriting. This invites further reflection on one of the questions with which this book began: the role of stories in climate transformations.

Both Onyesonwu and Phoenix make change possible through the ways in which they tell and re-tell their stories. Onyesonwu dictates her story in *Who Fears Death* to an unnamed Nuru scribe while she awaits her execution. She has come to realize that the conflict between the Okeke and Nuru is legitimized through ancient 'truths' codified in the Great Book.[160] However, there is a prophecy that a tall Nuru man with a long beard will rewrite the Great Book, and '[w]hen he comes, there will be good change for Nuru and Okeke'.[161] This

155 Diene (2022, pp. 101, 109).
156 Hanson (2016, p. 174). See also Chakrabarty (2012); Clark (2010, p. 141); Ghosh (2016, pp. 127–35); Johns-Putra (2018); Nixon (2011, p. 272).
157 Robinson (2019, p. 53). See Cole (2022, p. 140).
158 Mbembe (2019, p. 93).
159 Okorafor (2018, 2015a).
160 Okorafor (2018, p. 92). See also Burnett (2015, p. 142).
161 Okorafor (2018, p. 94).

prophecy is incorrect, however, and Onyesonwu begins to doubt other parts of the Great Book too. It is actually Onyesonwu who is destined to rewrite the Great Book, which she eventually manages through song. Onyesonwu retells how she rewrote the Great Book to the typist recording her story. 'My hand grew hot and I saw the symbols on my right hand split. The duplicates dribbled down onto the book where they settled between the other symbols into a script I still couldn't read. I could feel the book sucking from me, as a child does from its mother's breast.' This rewriting means that, as Onyesonwu says, '[t]he curse of the Okeke has been lifted. It never existed, *sha*.'[162] According to Burnett, the written word 'has a hybrid identity' for Okorafor: it is both 'the product and tool of the colonial oppressors and a potentially revolutionary tool in the postcolonial struggle against them.'[163]

This instability and unreliability of the written word, and the power of storytellers, is further emphasized in the prequel to *Who Fears Death*. *The Book of Phoenix* is told orally by Phoenix Okore and heard and transcribed by Sunuteel, an elderly man, out in the desert. Phoenix begins by saying: '[t]o tell my tale, I will use the old African tools of story: Spoken words. They are worthier of my trust and they will last longer. And during shadowy times, spoken words carry further than words typed, imaged or written.'[164] By her actions Phoenix changes the course of history, destroying western civilization in a fiery apocalypse and declaring, '[l]et that which had been written all be rewritten.' Sunuteel listens to Phoenix's story then twists it into the legend of the Okeke and Nuru which structures the genocide of *Who Fears Death*. Phoenix warns him: '"You can rewrite a story, but once it is written, it lives ... Who is writing you?" she asked. But her voice was fading.' His rewriting of her story causes pain and misery, 'until the one named Onyesonwu came and changed it again. But that is another story.'[165] Acts of writing and rewriting acquire political agency in these stories.

There is another interesting example of writing as political action in Mackay's *It Doesn't Have to Be This Way*. Luthando and Malcolm are software developers who write code for virtual reality worlds. Their professional work becomes activism when they launch a social media campaign showing climate breakdown under the hashtag #idhtbtw, standing for the book's title and central message, 'it doesn't have to be this way.'[166] Later in the story Malcolm sabotages his own trauma-deletion mental health app by writing and releasing

[162] Okorafor (2018, pp. 376–78).
[163] Burnett (2015, p. 146). See also Dowdall (2013, p. 12); Hoydis (2017); Pahl (2018).
[164] Okorafor (2015a, p. 6).
[165] Okorafor (2015a, pp. 221, 228, 232).
[166] Mackay (2022).

a virus to 'throw a spanner into the mindless, grinding cogs of progress'. He 'grins as he types the code', before releasing the malware that 'tears into the software and spreads'. Meanwhile, Luthando, on the outside of the dome, is 'putting together a library of stories, in case we don't survive this'. He is recording the lives, hopes, and dreams of ordinary people, to make 'a record of our lives and that makes this all seem a bit less bleak'. He stores them in caves on Table Mountain, illustrating the stories with rock art paintings on the walls.[167]

This concern with the power of words and symbols, writing, oral narratives, and song, runs through much Africanfuturist fiction. Eme and Udide compose music together in 'Spider the Artist'.[168] Dann is obsessed with recovering and recording the stories of the ancient lost civilizations that they discover in the library at the heart of The Centre, before they crumble into dust and sink into the marshes.[169] The power and longevity of stories is a central concern for Doris Lessing. In a speech at the award of her Nobel Laureate in 2007 she said: '[i]t is our stories that will recreate us, when we are torn, hurt, even destroyed. It is the storyteller, the dream-maker, the myth-maker, that is our phoenix, that represents us at our best, and at our most creative'.[170] In the *Mara and Dann* novels stories appear in fragments and memory is always partial and unreliable. There are ghostly echoes of ancient tales: Mam Bova, Ankrena, Mam Bedfly, Jull and Rom, Toski, and later Artemis and Actaeon, and Noah and the flood.[171] The epic tale ends with Griot—the storyteller—singing the songs of Mara and Dann. Little else will survive of them. David Sargeant concludes that Lessing's eco-political vision is that 'we cannot change our environmentally destructive forms of life, but the stories that are us can ... pass beyond it'.[172]

In the context of the histories and counter-histories of slavery and genocide, colonialism and racial capitalism, the question of who is able to write and tell these stories—and be heard—is a crucial one.[173] It is significant that so many of these stories feature strong black, women, often positioned as queer, or cyborg, or of outsider status, who not only survive the story to live into the future but who seize control of the narrative by writing and rewriting their own histories. Reading them suggests that rewriting the 'great books' and ecopolitical imaginaries of the present is a central challenge in order to avoid being haunted

[167] Mackay (2022).
[168] Okorafor (2011).
[169] Lessing (2006). See also Bryce (2019); Kotecki (2020); O'Connell (2016, p. 298); Pahl (2018).
[170] Lessing (2007). See also Sargeant (2016); Sperlinger (2017, p. 307); Watkins (2006, p. 258; 2010, pp. 150–64).
[171] Lessing (1999, pp. 166–7; 2006, p. 225).
[172] Sargeant (2016, p. 125).
[173] Haraway (2016); Weheliye (2014); Wynter and McKittrick (2015); Yusoff (2018).

by our own 'ghosts of the future'.[174] These novels can be read as part of a wider 'postmodern critique of narrative voice' in which hyper-reflexivity and suspicion of dominant and omniscient perspectives are slowly becoming more evident in climate fiction.[175] One of the most pointed political implication of these stories is that the eco-hero(in)es of the future will be those who seize the power to rewrite the story and narrate a new 'politicopoetic imaginary of the flesh'.[176]

Conclusion

In the spirit of this emphasis on the power of the storyteller I'd like you to imagine—as we approach the end of this book—how a long-term low-carbon political strategy might be written differently by a reader of Okorafor and Beukes, Onwualu and Lessing, Tsamaase and Mackay. To start with, the range of characters would be much wider. Rather than telling the story of how the state will drive top-down transformation, a diverse range of actors—including women, youth, queer, trans, differently abled, scientists, activists, farmers, traditional leaders, healers, scavengers, elders, religious leaders, and more-than-human entities—would have roles to play in low carbon transitions, tackling named enemies on the path to achieving a better future. Rather than being framed as an appeal for international financial support, these rewritten long-term strategies would articulate pathways to justice, restitution, and reparation, and to transformed structures of global power relations rather than just a transformed global energy system. By incorporating more-than-human perspectives, discourses of vulnerability, resilience, development, and growth would be complicated and rendered strange. These rewritten long-term strategies for low-carbon futures would be inhabited by believable people within meaningful communities and places, whose capacities for remembering the past, acting in the present, and imagining their own multiple futures would be integral to the story.

In this chapter I have shown how Africanfuturist climate fiction is animated by a revolutionary impulse to turn the world upside down. They lend weight to calls for climate reparations and show that by rewriting binaries—such as human/machine, human/nonhuman, freedom/domination, past/future, hero/villain, good/evil, normal/deviant, and reason/madness—we can imagine more capacious and inclusive subjectivities for climate-changed

174 Okorafor (2018, p. 187).
175 Johns-Putra (2018, p. 40).
176 Weheliye (2014, p. 132).

futures. Rather than provide detailed instructions for building post-capitalist, post-imperialist, post-fossil fuel societies—the 'specific floorplans' of better future worlds—these stories illuminate the unexamined assumptions, silences, and gaps of how climate futures tend to be narrated in contemporary global climate politics.[177]

Ultimately, fiction is a way to reflect on who we are and who we might become.[178] Climate fiction remains dominated by visions of the future where politics is shaped by white men, where the debts of empire, capital, and extraction remain unpaid. I have my own debts that reading these stories has brought into focus. Research involves intellectual debts, of course, but a career as an academic also relies upon sacrifices made by others: colleagues, families, students, communities, research subjects, and nonhumans too. I bear debts to society arising from intersectional structures of privilege. My consumption patterns and those of my family—including food, energy, and material commodities—entail ecological footprints and the exploitation of labour. Every word in the book is time and emotional attention taken away from my children, and yet they are inside this text too, in the gaps between words, sentences, chapters. My career as an academic is indebted to my parents, including their emotional support, career sacrifices, love of science fiction novels, and financial support arising not least from my father's life-long career with Exxon Chemicals, the implications of which creates its own debts: personal, familial, and corporate. Marco Grasso has calculated that ExxonMobil owes $549.6bn in cumulative reparations over the years 2025–2050 on the basis of their moral responsibility for climate harms knowingly perpetuated for profit.[179] Such debts, both individual and collective, are not easily discharged but must be lived with and reflected upon. They inform my research, activism, and teaching and they shape who I am.

Any act of writing is shaped by and in turn reshapes the debts we owe, the relationships we inhabit, and the futures we dream of. Writing this book has reinforced my belief that writing is politically significant in multiple ways, even if it is politically insufficient. Not many people are in the position to publish their view of the world, or even to have the space and time to write. We can all tell stories in some form, but the power to tell and retell stories that are listened to is a vital component of political agency. As Fanon proclaimed, '[m]astery of language affords remarkable power'.[180] But is this power enough in the face of the environmental crisis? As discussed in the Introduction to this book, Clark

177 Jameson (2007, p. 217).
178 Dauphinee (2010).
179 Grasso (2024, p. 741).
180 Fanon (2008, p. 9).

has questioned whether ecocriticism risks attributing an 'exaggerated sense of significant agency' to books and stories.[181]

Reading and telling tales is not an alternative to other forms of political action, nor do I think that storytelling should replace community organizing, political agitation, revolutionary struggle, or democratic engagement. Yet neither is it easy to separate 'the real business of political action' from texts and narratives.[182] Narratives structure our encounters with and reports of real politics, and even cultural commodities like books, short stories, and films are real, material interventions that can have effects beyond the cultural realm.[183] The key question, as Edkins suggests, is to ask what interventions particular stories enable. This is inherently subjective. As Edkins argues, '[t]here is no way that we can delineate with precision what might make some stories telling politically in a way other narratives are not. What is important is the practical effect they produce and what our judgement of that is. What they do is what counts—and how we judge that politically.'[184] From this perspective, imagination—including creativity, invention, fantasy, and speculation—are not supplements to politics, they are inextricable from politics.[185] As Grosz argues, '[p]olitics is an invention, a labour of fabrication, of experimentation with the unrepeatable and the singular, that links it more to intuition, to artistic production and aesthetic discernment than to planning, policy, or the extrapolation of existing relations.'[186]

As a result, I want to insist on the political necessity of reading and retelling Africanfuturist climate fiction. One of Okorafor's characters tells us that '[w]ords are powerful when chosen well and hurled with precision', and this is echoed by Mbembe: '[i]n the era of the Earth, we will effectively require a language that constantly bores, perforates, and digs like a gimlet, that *knows how to become a projectile*, a sort of full absolute, of will that ceaselessly gnaws at the real. Its function will not only be to force the locks but also to save life from the disaster lying in wait.'[187] Reading and writing different stories about alternative climate futures is an essential element of the political challenge of finding new ways of being and becoming human in the context of climate change.

[181] Clark (2015, p. 21).
[182] Edkins (2013, pp. 284–5).
[183] Daniel and Musgrave (2017); Grayson et al. (2009); Haiven and Khasnabish (2014); Iheka (2021); Jones and Parris (2018); Malm (2021, pp. 43–4); Milkoreit (2017, 2019); Selbin (2019).
[184] Edkins (2013, p. 292).
[185] Hammond (2021).
[186] Grosz (2004, p. 260).
[187] Mbembe (2019, p. 189, emphasis added); Okorafor (2015a, p. 201).

8
Rewriting the Ending

How will the story end? When the sun goes down? That's not an ending, just the start of the dark. As Frank Kermode observed, '[w]e cannot, of course, be denied an end; it is one of the great charms of books that they have to end.'[1] Adeline Johns-Putra has argued that much contemporary climate fiction seems animated by a wish for no ending at all, tending to refuse overarching resolutions. She concludes that in the era of the Anthropocene the 'happy-ever-after ending' is absurd, futile, and politically suspect.[2] Lessing seems to agree: Dann tells Mara's daughter, Tamar, that ancient stories sometimes ended, '[t]hey lived happily ever after', but he wonders 'what kind of people could say that? Or, perhaps they were just ... whistling in the dark, the way we do.'[3]

One of the weaknesses of the official long-term texts discussed in Chapter 2, 'Dominant narratives in long-term strategies', is that they fail to satisfactorily resolve the tensions within the narrative. They are unable to convincingly explain how 'we' get 'there', how obstacles will be overcome, and what future worlds might be like to live in. That is, perhaps, asking too much from these very specific types of policy texts. But advocates for long-term strategies have urged states to engage with deeper questions about the societies, politics, and economies of the future, and maybe policymakers and political visionaries can learn something from storytellers about how to craft a narrative that reaches a plausible, satisfying, and inspiring resolution.[4]

On the other hand, I sometimes worry that climate fiction stories are little more than whistling in the dark, as Dann feared. Certainly, the outpourings of novelists have not resolved the political contradictions of climate politics. The world is, at the time of writing, on track for at least 3°C of warming by the end of the century, which will have cataclysmic consequences for Africa and the world, humans and nonhumans.[5] The political responses are, as yet, woefully

[1] Kermode (2000, p. 23).
[2] Johns-Putra (2019, pp. 166–7); cf. Wallace-Wells (2019, p. 32).
[3] Lessing (2006, p. 280).
[4] Pathak (2017); Ross and Fransen (2017); Ross et al. (2021); UNDP (2017); Williams and Waisman (2017).
[5] Okereke (2024); Trisos et al. (2022).

African Climate Futures. Carl Death, Oxford University Press. © Carl Death (2025).
DOI: 10.1093/9780198960775.003.0008

inadequate, and African, American, European, and other governments continue to search for new fossil fuel resources to drill and burn. World-ending climate catastrophes have already arrived for many in the Global South and reparations for ecological debts are barely on the agenda.[6] Yet, even in this context, whistling in the dark is *not doing nothing*. Stories, like a tune whistled at night, can reassure and provide comfort, and surely we will need more of that in the difficult futures that lie ahead.[7] A whistle in the dark can also draw us closer to others, or transport us from our immediate circumstances into alternative imagined pasts, presents, or futures. Whistles can signal danger or summon a beloved human or nonhuman companion. Stories have many roles and considerable power. They are not a replacement for or alternative to political action, but they are both a manifestation of, and a representation of, political action. As Urry suggests, 'future visions have powerful social consequences'.[8] Imagined futures are a form of discursive power, a way to understand ourselves and others, and an encouragement to imagine things otherwise.

In this book I have argued that imagined futures are absolutely central to the global politics of climate change and that Africanfuturist climate fiction can fertilize the imagination of radical, reparative, more-than-human climate politics and policies. As Newell argues, '[n]ew futures have to be imagined for energy', as well as for transport, water, cities, agriculture, conservation, and every other part of society and the economy.[9] African socio-climatic imaginaries have tended to be marginalized, globally, but they are vital to consider in the struggle for just, equitable, and sustainable climate futures. While official political strategies and visions of the future deserve to be taken more seriously in political economies of transition, African long-term strategies exhibit shared silences in the ways that they envision what low carbon societies actually entail, how to get there, and the implications for democracy, subjectivity, and ecology. In contrast, the Africanfuturist climate fiction stories examined in this book imagine deeper and more radical visions of different spaces, temporalities, ecologies, and politics. Issues of land and climate justice, loss and damage, extractive political economies and the racialized and gendered violence of capitalism are more prominent in these stories than in most mainstream imaginaries. By reading the climate fiction of authors like Okorafor, Tsamaase,

[6] Patel and Moore (2017); Sultana (2022); Táíwò (2022).
[7] Alberro (2024, p. 2); Haiven and Khasnabish (2014, p. 20); Okorafor (2009, 2015b); Okri (2023); Thaler (2022, p. 272); Wallace-Wells (2019, p. 144).
[8] Urry (2016, p. 136).
[9] Newell (2021, p. 203).

Onwualu, Beukes, Lessing, and others as a contribution to debates in political theory, some of the limitations of mainstream narratives about climate futures can be highlighted.

Reiterating the storyline and summarizing the plot at the end of a long book are exhausting for the writer and the reader alike. Unlike readers of a fictional story, however, it is not unknown for readers of academic books to skip straight to the conclusion. For this reason, the next few pages will attempt to summarize the book's argument and structure, before focusing on the central political implications that emerge, in my reading, from Africanfuturist climate fiction.

In Chapter 2, 'Dominant Narratives in Long-term Strategies', I examine how African climate futures are produced by looking at some of the dominant political visions of future climate scenarios. Focussing on the official long-term strategies submitted by the governments of Nigeria, South Africa, Ethiopia, The Gambia, and Zimbabwe under the 2015 Paris Agreement of the UNFCCC, I show how they sought to narrate pathways for decarbonization, in line with goals of achieving net zero by mid-century. I set out four key sets of questions for these strategies: what are their plots and policies; what are the lead actors and supporting characters; what are the main discourses and themes; and what are some of the gaps and silences in these long-term strategies? There are some important similarities and differences that emerged from this analysis: they all portray the main agents of change as national governments and, with varying degrees of detail, they all position international financial support in a key supporting role. All the strategies begin by emphasizing the vulnerability of the country to the impacts of climate change. They all also emphasize their status as a developing country for whom a growth and poverty-alleviation focus is primary. They all at least mention the importance of a just transition; however, there are some important differences in the depth and scope of how this is narrated, and none of them articulate the case for climate reparations. Crucially, there are also important gaps in terms of the overarching visions of what a decarbonized economy would look like; how to get there, in terms of overcoming obstacles and enemies; the role of wider social engagement and democratic participation; the types of political subjects and agents imagined as driving the transition; and an ecological perspective that goes beyond anthropocentrism. Genuine political strategies for decarbonization need to be both more imaginatively ambitious and more honest about the balance of forces in the struggle for a just transition. The scope and ambition of these long-term strategies represent a political opportunity to explore more radical and diverse narratives of imagined future climates, and

I argue that interrogating their strengths, weaknesses, and silences is a crucial task for the political economy of energy transitions.

The rest of the book takes up this challenge, arguing that Africanfuturist climate fiction offers some powerful contributions to re-envisioning ecopolitical imaginaries. In Chapter 3, 'Why Read African Climate Fiction?' I begin by asking what do stories *do*, politically? I argue that climate fiction can play important roles in how we understand our pasts, diagnose our presents, and in opening up imaginative possibilities for our futures. Reading, writing, and discussing fiction is an essential element of how we understand ourselves and others. After making the case for the political implications of reading fiction, Chapter 3 provides a brief introduction to the burgeoning field of climate fiction more generally, before exploring the role of stories from and about Africa in this field. In contrast to the historic and recent prominence of African stories and authors in fictional literature more generally, African plots and authors remain relatively marginal in science fiction and climate fiction. Yet, there is a growing body of Africanfuturist climate fiction and this chapter introduces the authors that the rest of the book discusses in more depth. They include Nnedi Okorafor, Doris Lessing, Wanuri Kahiu, Lauren Beukes, Terh Agbedeh, Suyi Davies Okungbowa, Chinelo Onwualu, Tlotlo Tsamaase, Alastair Mackay, Jessica Wilson, and Vuyokazi Ngemntu. These authors, I argue, comprise a body of Africanfuturist climate fiction that has emerged from much older traditions of African self-writing, speculative fiction, and anticolonial Africanfuturism.

In Chapter 4, 'Strange Places and Unfamiliar Homesteads', I ask where these stories are set. African landscapes are rarely prominent in more dominant climate imaginaries, except as an apocalyptic warning of dystopian futures of deserts, slums, and warzones. By reading these Africanfuturist stories we can visit a much more diverse set of story-worlds: from the gleaming towers and monorails of futuristic Abuja and Kinshasa, to Johannesburg's mine dumps and space-age East African compounds; from homely eco-villages in the Niger Delta, to hospitable deserts from the Sahara to the Kalahari, or the flooded nightmares of dystopian Lagos or Cape Town. In contrast to eco-fiction that relies on invoking familiar and much-loved places, these Africanfuturist stories are often located in unsettling, unfamiliar landscapes that can be described as heterotopian in the way that they disrupt and disturb our maps of the world. Moreover, while many of them have an urban focus they also work to destabilize utopian and hyper-modernist 'African urban renaissance' storylines. They emphasize the political necessity of flight and exile from cities and also warn us that families and homes can be places of danger and insecurity. We need to imagine cities that are fluid, porous, and in-flux, and domestic environments and care-relations that are imaginative and hospitable.

In Chapter 5, 'Heterotopian Temporalities and Afrocentric Futures', I reflect upon when these stories are set and their implications for how we think about time. There are four main ways in which these stories challenge dominant linear and cyclical temporalities. First, they imagine Afrocentric global futures in which the African continent is at the centre of global politics. This turns conventional assumptions about Africa and 'progress' on their head. Second, they envision plural and diverse temporalities that interweave past, present, and future in multiple ways. In these stories time loops and spirals, there are multiple co-existing 'nows' and futures that reshape the past. Third, they insist on the active role of the detritus of the past—including trauma and ghosts—in the present and future. Colonialism, racism, and Whiteness continue to structure the present and the future. Paradoxically, this uncanny persistence of the past is woven into a utopian, revolutionary politics, urging us to imagine brave new worlds.[10] Finally, this is all predicated on the inevitability of death, decay and suffering. The mortality of our bodies—and our civilizations—is a fundamental premise of socio-climatic futures. As such, climate strategies need to reflect on how to preserve, memorialize and repair that which has been bequeathed to us.

In Chapter 6, 'Dangerous Ecologies and More-than-Human Conviviality', I focus on who these stories are about, paying particular attention to relations between human and nonhuman characters. I argued that while dominant narratives about climate futures are typically anthropocentric, Africanfuturist climate fiction can make an important contribution to reimagining relations between humans and animals, forests, deserts, and waters. I show, first, how questions of how to relate well and engage ethically with nonhuman others are prominent in most, if not all, the stories examined in this book. Second, in many of the Africanfuturist stories examined here the imagined new relationships entail the emergence of quite radically new forms of human subjectivity. Third, these changing relationships and changing subjectivities produce new visions of living with very different others. Angry, predatory, homicidal species and environments are much more prominent in Africanfuturist climate fiction—crocodiles and vultures, dangerous waters, and inhospitable desert landscapes—than the more familiar companion species of polar bears, butterflies, albatross, and trout that have become iconic in climate activism and fiction. Africanfuturist fiction is profoundly convivial, concerned with ways in

[10] Kelley (2002); Womack (2013); Zamalin (2019).

which a more-than-human 'we' might rub along together in the future, but also insistently cognisant of the trouble that this will bring.

In Chapter 7, 'Reparative climate politics and reimagined subjectivities', I ask more directly what sort of politics these stories imply. Most Africanfuturist climate fiction is radical and revolutionary, committed to imagining how to overturn the existing order of things. The second part of the chapter considers the question: what comes next? There are powerful reparatory themes in many of these novels, lending emotive weight to calls for climate reparations and the healing of loss and damage.[11] As Jackson concludes, art can hold out the possibility of redress or remedy, and 'may be a means of setting right a wrong'.[12] These stories offer hyper-modernist visions of science and technology-driven rampant capitalism, neo-feudal tales of warlords and chieftains, and a few glimpses of post-capitalist political economies. However, Africanfuturist climate fiction makes an even stronger contribution to climate justice politics in terms of the reimagination of human subjectivity. Many of these stories feature marginal, queer, cyborg, feminist, outcast heroines, posthuman actors, and radical black utopians. By imagining more 'accessible futures' in which a more diverse range of eco-heroes and climate-changed characters live, they can add vitality to policy narratives about socio-climatic futures.[13] Significantly, a repeated form of political agency in these stories is the power to write and rewrite stories themselves. Those whose control the story, and can alter the endings, have the power to shape and reshape climate politics.

Political Implications

At the end of this particular story, therefore, we can now ask: what lessons might Africanfuturist climate fiction have to teach those who are drafting policy strategies, negotiating agreements, shaping investment decisions, organizing in communities, marching on the streets, teaching in schools and universities or writing new stories about the future? In the conclusion to the previous chapter I imagine how a long-term low carbon political strategy might be written differently, inspired by Africanfuturist climate fiction. Here, I draw together and make explicit the political implications explored throughout the book. These suggestions are provocations for further research and praxis, rather than a new blueprint for climate policy and activism.

[11] Táíwò (2022).
[12] Jackson (2020, p. 214).
[13] Kafer (2013).

Primarily, and most obviously, this book argues that stories are important. How we imagine the future is important. Political actors and policymakers need to pay more attention to the stories they tell—and those they are reading or listening to—and actively seek to diversify and radicalize how the future is narrated. Fictional stories can enrich the ways in which political strategies are narrated and authors and storytellers should be regarded as important actors in governing future climates. We teach our children through storytelling and they teach us in turn. Climate stories should be central to how climate change is taught and researched, negotiated, and governed. Moreover, if we want to survive to tell new stories in liveable futures then we need to transform carboniferous capitalism, and soon.

Africanfuturist climate fiction stories remind us that Africa and Africans are not marginal to climate futures. Often described as being on the frontlines of climate politics, African futures will be central to global climate futures and recognizing this means we need to tell stories about climate justice, revolution, reparation, and restitution, and identity, memory, and trauma. We need to tell—and listen to—stories that interrogate racism, Whiteness, patriarchy, homophobia, ableism, and speciesism much more prominently than in mainstream climate politics or the mainstream climate fiction canon. Africans have a place in climate-changed futures and reflecting on why stories from and about Africa have been so often rendered marginal can illuminate continued forms of coloniality in climate politics.

While so much environmental activism relies upon appeals to the landscapes and species that are familiar and loved, these stories suggest that we can't remain here. The landscapes of the future will be strange, different, heterotopian spaces that disrupt and disturb our maps of the world and will require new ways of being and relating to place. Through the imagination of unfamiliar people and places, climate fiction can help us view our own environments more critically. Stories can help us visit places we have never been to and to reimagine places we think we know. Environmental justice will require hybrid interventions and a willingness to engage with the uncanny and eerie, rather than just a rear-guard defence of essentialist localism. Climate justice struggles will not only be for and over familiar and loved landscapes—our backyards, national parks, bucolic rural farmlands, and garden cities—but will take place within and over strange and unsettling spaces and places. New ways of protecting, conserving and interacting with these spaces may be needed.

These stories suggest that policy-makers, activists, and communities should beware techno-modernist fantasies of smart, eco-cities. The hyper-modernist dreams of engineers and architects tend to work in the interests of the rich and powerful and produce new forms of exclusion and injustice. Urban politics is intimately connected to peri-urban and rural hinterlands, and the ways in which people move between cities and surrounding areas—the routes of escape from the city—are crucial for climate politics. We need to devise political strategies and ecopolitical imaginaries for cities that are fluid, porous, hospitable, and in-flux.

Fiction can also bring the everyday, the mundane, and the personal or private sphere into much clearer and richer focus. Rather than simplistically invoking the importance of families, or households, or parental love, for climate politics, these stories suggest that families are relations that are changing and should and will change in climate-changed futures. The ethics and political economy of care will be absolutely central in the societies of the future, but rather than romanticizing traditionally feminine, heteronormative care roles, or nuclear family bonds, critical researchers should politicize them. Many of the stories examined in this book highlight the disruptive impact of gestation and birth, which can work to unsettle and re-politicize the utopian imagination and contribute to the challenge of imagining sustainable future spaces, and places 'in which we would like to live, work, love, and create'.[14] We need climate-changed domestic environments and care-relations that are diverse, imaginative, and hospitable.

The re-imagination of plural and non-linear timelines and temporalities in these stories offers lessons to those who draft political strategies, scientific forecasts, business scenarios, and social movement campaigns. Rather than projecting a straightforward pathway from here to there, incorporating multiple co-existing 'nows' and futures that reshape the past can be a way to bring these narratives to life and engage wider publics in democratic debate and deliberation. This must go beyond the sketching out of different potential scenarios or emissions trajectories, to a fuller reflection on diverse temporalities within different areas of transition politics, such as energy, transport, agriculture, the lived environment, and culture and heritage. Different individuals and social groups—including different generations, cultures, languages, regions, and classes—may well have differing and overlapping temporal worldviews that could shape their perspectives on the direction and pace of transition pathways.

Another way of thinking about the political implications of temporality in Africanfuturist climate fiction is to consider how to preserve, memorialize, and repair that which has been bequeathed to us. Academic researchers on

[14] Zamalin (2019, p. 139).

transitions as well as policy-makers need to pay more attention to the traces of the past in the present and future, and ask: how will the museums and histories of the future narrate our present? What is the role of memory, tradition and even death, within a radical, progressive politics of climate justice and reparation? What should be preserved and what should be transformed or rejected? In terms of the politics of the UNFCCC negotiations, these stories lend weight to demands for significantly expanding the compensation paid by the Global North to the Global South through institutions like the Loss and Damage Fund, but they also warn that money itself is not enough. World-remaking reparations require cultural critique, a shift in values and power-relations and pluralizing of imagined futures. What debts do we each owe, individually and collectively, and how might these be repaid? There should be more space for remembrance, grief, mortality, and compost in dominant climate imaginaries. It is here that Africanfuturist climate fiction can play an important role in narrating and disseminating new ecopolitical imaginaries.

Another implication that comes from these stories is the vivid reminder that we are not the only species to be facing climate-changed futures. Climate fiction is a powerful amplifier of arguments that more-than-human politics will involve changing our relationships with other species of plant and animal, as well as the land, air, and water.[15] More specifically, these stories suggest that attempts to foster better ethical relations with nonhumans may well involve danger and sacrifices in some form, and that human actors need to attempt to make multispecies reparations for harms already done. This also means going beyond imaginations dominated by polar bears, butterflies, trout, and pandas. These stories make it very clear that a post-anthropocentric politics means engaging with—and living and dying well alongside—the crocodile, vulture, spider, bedbug, bacteria, and virus. Exactly what this means in practice is hard to imagine, but there is an extensive body of creative research, such as on inter-species diplomacy, that can help activists and policymakers do just that.[16] Reflecting on our own lived experiences—and the more-than-human encounters that constitute our homes and workplaces—is an important site for envisioning better climate futures.

As well as challenging the anthropocentrism of climate politics, these stories also challenge how researchers and theorists imagine human agency and how we understand humanity itself. Africanfuturist stories about posthuman shape-shifting can teach us that embracing human difference—racial,

[15] Iheka (2017, 2021).

[16] Cudworth and Hobden (2022); Liboiron (2021); Mitchell (2016); Sze (2020); Tschakert (2022); Youatt (2014).

gendered, sexual, differently-abled, neuro-diverse—will be vital in building accessible futures as the climate changes. These stories diversify the imagined subjects of climate-changed futures, narrating queer, cyborg, outcast subjects as central to the politics of climate futures, although not as heroic individuals but within more-than-human planetary entanglements. We need to imagine more capacious and inclusive subjectivities for climate-changed futures. For Gibson, 'privileged futurities must "give way" in both the texts themselves and the world beyond the page/screen if marginalized futurities are to proliferate'.[17] Acknowledging this has multiple implications: for the diversity and inclusivity of climate movements, organizations, institutions, spaces, and discourses; for the ways in which political agency is imagined and represented; and for the strategy and planning of climate transitions in areas like education, culture, technology, transport, cities, and so on.

Reading climate fiction has also made clearer some of the limitations of this methodological approach to crafting political theory and socio-climatic imaginaries fit for the challenges of climate-changed futures. As discussed in the introduction, there is a risk of cultural appropriation at the heart of this book project. Yet, I think there is a greater risk of failing to listen to stories from different cultural contexts, of failing to reflect upon them, learn from them, challenge them, and be open to how they challenge us. Moreover, of course, fictional narratives do not have all the answers. Fiction can also obscure or omit issues, just like other political imaginaries. In many of these stories there is little sustained exploration of climate science or new technologies, and not much engagement with the institutional politics of governing future climates. There is also the risk in fictional narratives that individual agency and lone heroes are over-emphasized. It is important to explicitly reiterate that a fictional story is not, itself, usually a work of theory, in the sense of a contribution to a tradition of academic debate. Climate theory can be inspired and strengthened by drawing on fictional stories—and artists, activists, politicians, and academics can all play a role in this—but a fictional story and a contribution to academic climate theory are written and judged differently. Importantly, the function of artists and creatives should not be reduced to an instrumental role of providing new ideas, visions, and imaginaries to enrich political and policy deliberations. Rather, the burden is on political leaders, activists, and intellectuals to construct and articulate radical, popular, sustainable, coherent, and progressive imagined visions of the future.

[17] Gibson (2020, p. 216).

Ultimately, these stories insist that we take seriously the political power of stories and narratives themselves. These are radical stories, ones that encourage us to persist, to struggle, and confront the powers that be. They can be read as a call to take up the fight against the status quo, in many different ways, and fortify activists for the ongoing battle against the forces of capital, empire, and Whiteness.[18] But the role of stories—as fertile texts for insurgent imaginaries—is not simply as a resource or tool. As Grosz suggests, political struggle 'is directed, at its best, to a future we cannot directly recognize, a future that does not simply extend our current needs and wants but may actively transform them in ways we may not understand or control.'[19] Radically imaginative and disturbing stories are needed to stimulate critical self-reflection and defamiliarization. Challenging dominant or harmful stories, rewriting endings, telling alternative stories: these are some of the most important political sites for reflecting critically on contemporary climate politics. They are part of, and indeed inseparable from, the narrative representations that constitute other political sites like COP summits, shareholder meetings, party conferences, social movement assemblies, classrooms, government offices, factory floors, picket lines, and households.[20] Stories emerge from these sites of political contestation, and stories shape and reproduce the identities, discourses, and socio-climatic imaginaries that constitute these political spaces.

* * *

There is a particular challenge faced when trying to end a book on stories and imagined futures. A good ending is one of the characteristics of great fiction, whereas endings in social science texts are often more perfunctory. Maybe they are given less thought. Maybe one of the distinctive characteristics of a fictional story is that it has a start, middle, and end, in a way that 'real life', or social scientific analysis of what purports to be real life, does not. I quoted Kermode's remark above, that it is one of the great charms of books that they have to end. This could be read somewhat flippantly, in the sense that it is a relief to get to the end of something and move on to another task. But there is something going on here that is more deeply connected to an essential attribute of a fictional narrative. As Kermode goes on to explain, 'to make sense of our lives from where we are, as it were, stranded in the middle, we need fictions of beginnings and fictions of ends, fictions which unite beginning and end and

[18] Haiven (2014); Imarisha and Brown (2015); Kelley (2002); Malm (2021).
[19] Grosz (2004, p. 260).
[20] Bleiker (2009); Haiven and Khasnabish (2014); Patel and Moore (2017, p. 211); Sultana (2022).

endow the interval between them with meaning.'[21] A fictional story imposes an ending—and hence some kind of meaning and closure, even if partial and opaque—onto a representation of life that would otherwise lack it.

There are stories that I've loved that I'd struggle to tell you how they end, and there are some stories whose endings stay with me, years after I have read them. Octavia E. Butler's *Parable of the Sower*—the first volume of an unfinished series—ends with Olamina and her companions burying their dead and planting oak trees. 'Afterward, we sat together and talked and ate a meal and decided to call this place Acorn.'[22] Samuel R. Delany ends *Stars in My Pocket like Grains of Sand* with Marq Dyeth ruminating on the slipperiness of the concept 'morning' in the context of interstellar travel and planets with different solar orbits. Falling back into sleep on a colony-ship, they realize 'that morning, whatever it was, was over. When I awoke again, it would be day.'[23]

I often feel a deep and pleasurable sadness in finishing a good story, saying goodbye to characters in whose company many hours have passed. A favourite ending of mine is the dark optimism of Pratchett and Gaiman's novel, *Good Omens*. 'If you want to imagine the future, imagine a boy and his dog and his friends. And a summer that never ends . . . imagine a sneaker, laces trailing, kicking a pebble; imagine a stick, to poke at interesting things . . . imagine a figure, half angel, half devil, all human . . . Slouching hopefully towards Tadfield. For ever.'[24] I read a lot of Terry Pratchett stories when I was young and now that I am surrounded by young children exploring and imagining their own worlds into being, traversing and gradually leaving their childhood, I feel the nostalgic lure of the child-who-never-grows-up, the summer that never ends, as an imaginary suffused with irreducible hope in humanity. Part of the reason I am so drawn to fiction at this point in my life may well be the capacity of stories to freeze time: you can always go back and read a story again. And yet . . . you never read a story the same way twice, as it's not the same story and you're not the same person. As I finished Doris Lessing's *Mara and Dann* stories for a second time I wondered if I'd ever read them again. Probably not, and I grieve gently over my self-imposed exile from them. These stories have been my close companions over the past few years (see Figure 7). Writing this book has been a way to extend my time in their world.

Many of the most interesting and powerful endings in the Africanfuturist climate fiction stories discussed in this book have already been recounted in

[21] Kermode (2000, p. 190). See also White (2011).
[22] Butler (2019, p. 311).
[23] Delany (2004, p. 347).
[24] Pratchett and Gaiman (2014, p. 402).

Figure 7 Photo taken by author

previous chapters. Asha lies dead or dying on the desert soil in Kahiu's *Pumzi*, sheltering the Maitu seed as the camera pans out to reveal a vast forest. The spider-narrator, Udide Okwanka, decides to leave his web and become part of the story at the close of Okorafor's *Lagoon*. AO turns off the turbines and ends the Red Eye storm at the culmination of *Noor*. Lessing has Dann and Griot reflect on the stupidity of mankind as her epic *Mara and Dann* novels conclude. Milo feels himself becoming plant-like in a cave on Table Mountain in Mackay's *It Doesn't Have to Be This Way*. At the end of Beukes' *Zoo City*, Zinzi heads north across the Beit border bridge, over crocodile-infested waters, to get lost for a while.

The most powerful ending, however, comes in Okorafor's *Who Fears Death*. The story has been overshadowed throughout by Onyesonwu's death foretold; being shown her own personal ending is a condition of becoming a sorcerer.[25] The novel acquires a more intense mythic tone as it progresses, inexorably, towards the blood-spattered culmination. As envisioned, after rewriting the Great Book and saving the people of the Seven Rivers Kingdom, Onyesonwu

[25] Okorafor (2018, p. 155).

is stoned to death, then cremated, and buried in the desert. There is a double rewriting that takes place, however, as no sooner has Okorafor finished Onyesonwu's tale then she rewrites it, explicitly disrupting and twisting the linear structure of the novel itself. The epilogue is not the end nor is Onyesonwu's predestined death the last word. The last chapter is entitled 'Chapter 1: Rewritten'. Rather than be taken from the cell to her death, Onyesonwu changes into a mythical creature, a fire-spitting *Kponyungo*, immolates her captors, and she 'blew into the sky like a shooting star, ready to return home'. Perhaps, her mentor reflects, 'she'd want to *live* in the very world she helped to remake?'[26]

In life, of course, there are no clear endings. And yet there are worlds ending all around us.[27] In terms of climate change politics, Bill McKibben has warned that '[t]he end of this story is written, unless we rewrite it. Fast.'[28] The final lesson I take from these climate fiction stories is from Onyesonwu: we need to try to live in the worlds that we've imagined. We can intervene politically to enact the best visions of these stories or rewrite their endings to craft new stories. These stories—in themselves, as texts—are not especially significant, politically. It is reading them in a particular way, discussing them, critiquing them, and other ways of enacting them that renders them political. Alexandria Ocasio-Cortez insists that 'our future has not been written yet'; rather, I would say that rewriting our futures is always possible.[29] Join a march, stand on a picket, organize a workplace or community, oppose drilling or logging or other forms of extraction, teach a classroom, reflect on how to repay your debts. Then start telling stories about what you've done or the futures you have envisioned. I want to hear these stories and I imagine I'm not alone.

[26] Okorafor (2018, pp. 383–5)
[27] Yusoff (2018, pp. 12–13).
[28] McKibben (2013, p. 747).
[29] Klein (2019).

Bibliography

Abrahamsen, Rita (2003) 'African Studies and the Postcolonial Challenge', *African Affairs*, 102, 407, pp. 189–210.

Abrahamsen, Rita, Barbra Chimhandamba, and Farai Chipato (2023) 'Introduction: The African Union, Pan-Africanism, and the Liberal World (Dis)Order', *Global Studies Quarterly*, 3, pp. 1–10.

Abram, Nia (2018) 'Wakanda Forever: Anticolonialism as Sustainable Technology', *Terra Incognita*, 3 April 2018. Available at https://www.terraincognitamedia.com/features/wakanda-forever-anticolonialism-as-sustainable-technology2018 (accessed 23 November 2023).

Achebe, Chinua (1958) *Things Fall Apart* (London: Heinemann).

Adam, Barbara and Chris Groves (2007) *Future Matters: Action, Knowledge, Ethics* (Leiden: Brill).

Adejunmobi, Moradewun (2016) 'Introduction: African Science Fiction', *Cambridge Journal of Postcolonial Literary Inquiry*, 3, 3, pp. 265–72.

Adesida, Olugbenga (1994) 'Futures Studies in Africa', *Futures*, 26, 9, pp. 884–90.

African Energy Commission (2019) *Designing the African Energy Transition: An approach for Social and Economic Transformation in a Climate Compatible Manner* (Algiers, AFREC).

African Union (2015) *Agenda 2063: The Africa We Want* (Addis Ababa: African Union Commission).

Agathangelou, Anna M. and Kyle D. Killian (2016) 'International Relations as a Vulnerable Space: A Conversation with Fanon and Hartman about Temporality and Violence', in Anna M. Agathangelou and Kyle D. Killian (eds), *Time, Temporality and Violence in International Relations: (De)Fatalizing the Present, Forging Radical Alternatives* (Abingdon: Routledge), pp. 23–42.

Agbedeh, Terh (2013) 'Mango Republic', in Ayodele Arigbabu (ed.), *Lagos_2060: Exciting Sci-fi Stories from Nigeria* (Lagos: Dada Books), pp. 180–99.

Aiyetoro, Mary Bosede and Elizabeth Olubukola Olaoye (2016) 'Afro–Science Fiction: A Study of Nnedi Okorafor's *What Sunny Saw in the Flames* and *Lagoon*', *Pivot*, 5, 1, pp. 226–46.

Ajibade, Idowu, Mark Pelling, Julius Agboola, and Matthias Garschagen (2017) 'Sustainability Transitions: Exploring Risk Management and the Future of Adaptation in the Megacity of Lagos', *Journal of Extreme Events*, 3, 3, pp. 1–25.

Akinsiku, Barakat (2021) 'The God of the Sea', in Angie Dell and Joey Eschrich (eds), *Everything Change: An Anthology of Climate Fiction*, Volume III (Arizona: Center for Science and the Imagination), pp. 19–31.

Alami, Ilias, Adam D. Dixon, and Emma Mawdsley (2021) 'State Capitalism and the New Global D/development Regime', *Antipode*, 53, pp. 1294–318.

Alberro, Heather (2024) *Terrestrial Ecotopias: Multispecies Flourishing in and Beyond the Capitalocene* (Oxford: Peter Lang).

Albert, Michael J. (2024) *Navigating the Polycrisis: Mapping the Futures of Capitalism and the Earth* (Cambridge MA: MIT Press).

Allers, Roger and Rob Minkoff (1994) *The Lion King* [film] (Walt Disney Films).

Andrews, Kehinde (2024) 'The Knife Is Still in Our Backs: Reparations Washing and the Limits of Reparatory Justice Campaigns', *Development and Change*, 55, 4, pp. 628–50.

Arigbabu, Ayodele (ed.) (2013) *Lagos_2060: Exciting Sci-fi Stories from Nigeria* (Lagos: Dada Books).

Arseneault, Jesse (2019) 'African Fictions, Animal Figures and Anthropocentric Frameworks', in Moradewun Adejunmobi and Carli Coetzee (eds), *Routledge Handbook of African Literature* (Abingdon: Routledge), pp. 231–43.

Ashcroft, Bill (2017) *Utopianism in Postcolonial Literatures* (Abingdon: Routledge).

Atwood, Margaret (1985) *The Handmaid's Tale* (Toronto: McClelland and Stewart).

Azzarello, Robert (2016) *Queer Environmentality: Ecology, Evolution, and Sexuality in American Literature* (London: Routledge).

Baek, Yong Jun, Tae Yong Jung, and Sung Jin Kang (2019) 'Low Carbon Scenarios and Policies for the Power Sector in Botswana', *Climate Policy*, 19, 2, pp. 219–30.

Bai, Xuemei, Sander van der Leeuw, Karen O'Brien, Frans Berkhout, Frank Biermann, Eduardo S. Brondizio, Christophe Cudennec, John Dearing, Anantha Duraiappah, Marion Glaser, Andrew Revkin, Will Steffen, and James Syvitski (2016) 'Plausible and Desirable Futures in the Anthropocene: A New Research Agenda', *Global Environmental Change*, 39, pp. 351–62.

Baker, Lucy, Jesse Burton, and Hilton Trollip (2021) 'The Energy Politics of South Africa', in Kathleen Hancock and Juliann Allison (eds), *The Oxford Handbook of Energy Politics* (Oxford: Oxford University Press), pp. 711–29.

Baldwin, Andrew (2012) 'Whiteness and Futurity: Towards a Research Agenda', *Progress in Human Geography*, 36, 2, pp. 172–87.

Baker, Lucy, Peter Newell, and Jon Phillips (2014) 'The Political Economy of Energy Transitions: The Case of South Africa', *New Political Economy*, 19, 6, pp. 791–818.

Ballard, James G. (1962) *The Drowned World* (New York: Berkley Books).

Barnett, David (2023) 'It's Not Climate Change, It's Everything Change': Sci-fi Authors Take on the Global Crisis', *The Guardian* (UK), 9 July 2023. Available at https://www.theguardian.com/books/2023/jul/09/climate-change-sci-fi-authors-global-crisis-atwood-mccarthy (accessed 8 March 2024).

Bassey, Nnimmo (2012) *To Cook a Continent: Destructive Extraction and the Climate Crisis in Africa* (Cape Town: Pambazuka Press).

BBC (2013) *Africa* [television documentary series] (BBC).

Beck, Marisa (2018) 'Telling Stories with Models and Making Policy with Stories: An Exploration', *Climate Policy*, 18, 7, pp. 928–41.

Becket, Fiona (2009) 'Environmental Fables? The Eco-Politics of Doris Lessing's "Ifrik" Novels', in Alice Ridout and Susan Watkins (eds), *Doris Lessing: Border Crossings* (London: Continuum), pp. 129–41.

Beckert, Jens (2016) *Imagined Futures: Fictional Expectations and Capitalist Dynamics* (Cambridge, MA: Harvard University Press).

Bell, Duncan (2021) 'J. G. Ballard's Surrealist Liberalism', *Political Theory*, 49, 6, pp. 934–67.

Bennett, Elena M., Martin Solan, Reinette Biggs, Timon McPhearson, Albert V. Norström, Per Olsson, Laura Pereira, Garry D. Peterson, Ciara Raudsepp-Hearne, Frank Biermann, Stephen R. Carpenter, Erle C. Ellis, Tanja Hichert, Victor Galaz, Myanna Lahsen, Manjana Milkoreit, Berta Martin López, Kimberly A. Nicholas, Rika Preiser, Gaia Vince, Joost M. Vervoort, and Jianchu Xu (2016) 'Bright Spots: Seeds of a Good Anthropocene', *Frontiers in Ecology and the Environment*, 14, 8, pp. 441–8.

Bennett, Jane (2018) '"Queer/White" in South Africa? A Troubling Oxymoron', in Zethu Matebeni, Surya Monro, and Vasu Reddy (eds), *Queer in Africa: LGBTQI Identities, Citizenship, and Activism* (Abingdon: Routledge), pp. 99–113.

Bertelsen, Eve (1986) 'Interview with Doris Lessing', *Journal of Commonwealth Literature*, 21, 1, pp. 134–61.

Bethlehem, Louise (2015) 'Scratching the Surface: The Home and the Haptic in Lauren Beukes's *Zoo City* and Elsewhere', *Scrutiny2*, 20, 1, pp. 3–23.

Beukes, Lauren (2010) *Zoo City* (New York: Mulholland Books).

Beukes, Lauren (2017) 'The Power of Afrofuturism', *Nature*, 552, p. 330.

Bina, Olivia, Andy Inch, and Lavínia Pereira (2020) 'Beyond Techno-utopia and Its Discontents: On the Role of Utopianism and Speculative Fiction in Shaping Alternatives to the Smart City Imaginary', *Futures*, 115, pp. 1–14.

Bleiker, Roland (2001) 'The Aesthetic Turn in International Political Theory', *Millennium*, 30, 3, pp. 509–33.

Bleiker, Roland (2009) *Aesthetics and World Politics* (London: Palgrave Macmillan).

Blomkamp, Neill (2009) *District 9* [film] (Sony Pictures).

Bollinger, Laurel (2007) 'Placental Economy: Octavia Butler, Luce Irigaray, and Speculative Subjectivity', *Literature Interpretation Theory*, 18, 4, pp. 325–52.

Boswell, Parley Ann (2014) *Pregnancy in Literature and Film* (Jefferson: McFarland & Company).

Bottici, Chiara and Benoît Challand (eds) (2011) *The Politics of Imagination* (Abingdon: Birkbeck Law Press).

Bould, Mark (2009) 'Introduction: Rough Guide to a Lonely Planet, from Nemo to Neo', in Mark Bould and China Miéville (eds), *Red Planets: Marxism and Science Fiction* (London: Pluto), pp. 1–26.

Bould, Mark (2015) 'African Science Fiction 101', *SFRA Review*, 311, pp. 11–18.

Bowman, Andrew (2020) 'Parastatals and Economic Transformation in South Africa: The Political Economy of the Eskom Crisis', *African Affairs*, 119, 476, pp. 395–431.

Boykoff, Maxwell and Beth Osnes (2019) 'A Laughing Matter? Confronting Climate Change through Humor', *Political Geography*, 68, pp. 154–63.

Brady, Amy (2019) 'The Uncanniness of Climate Change', Burning Worlds Column, *The Chicago Review of Books*, 18 September. Available at https://chireviewofbooks.com/2019/09/18/the-uncanniness-of-climate-change/ (accessed 8 March 2024).

Brady, Amy, Carolyn Kormann, Pitchaya Sudbanthad, Eve Mosher, and Robert Moore (2019) 'Four Storytellers Tackle Climate Change', *Guernica*, 15 October. Available at https://www.guernicamag.com/four-storytellers-tackle-climate-change/ (accessed 26 October 2019).

Branch, Adam and Zachariah Mampilly (2015) *Africa Uprising: Popular Protest and Political Change* (London: Zed Books).

Braun, Rebecca (2022) 'Literary Futures: How Fiction Can Help Policy Makers', in Carlos López Galviz and Emily Spiers (eds), *Routledge Handbook of Social Futures* (London: Routledge), pp. 189–97.

Breznican, Anthony (2017) 'How Black Panther Solves the Problem of M'Baku', *Entertainment Weekly*, 13 July. Available at https://ew.com/comic-con/2017/07/13/black-panther-mbaku-man-ape-issue/ (accessed 2 August 2023).

Bridge, Gavin, Stefan Bouzarovski, Michael Bradshaw, and Nick Eyre (2013) 'Geographies of Energy Transition: Space, Place and the Low-Carbon Economy', *Energy Policy*, 53, pp. 331–40.

Brooks, Lonny J. Avi, Jason Tester, Eli Kosminsky, and Anthony D. Weeks (2022) 'Queering: Liberation Futures with Afrofuturism', in Carlos López Galviz and Emily Spiers (eds), *Routledge Handbook of Social Futures* (London: Routledge), pp. 260–70.

Bryce, Jane (2019) 'African Futurism: Speculative Fictions and "Rewriting the Great Book"', *Research in African Literatures*, 50, 1, pp. 1–19.

Burke, Anthony (2023) 'Interspecies Cosmopolitanism: Non-human Power and the Grounds of World Order in the Anthropocene', *Review of International Studies*, 49, 2, pp. 201–22.

Burnett, Joshua Yu (2015) 'The Great Change and the Great Book: Nnedi Okorafor's Postcolonial, Post-Apocalyptic Africa and the Promise of Black Speculative Fiction', *Research in African Literatures*, 46, 4, pp. 133–50.

Büscher, Bram and Robert Fletcher (2020) *The Conservation Revolution: Radical Ideas for Saving Nature beyond the Anthropocene* (London: Verso).

Butler, Cameron (2017) 'A Fruitless Endeavour: Confronting the Heteronormativity of Environmentalism', in Sherilyn MacGregor (ed.), *Routledge Handbook of Gender and Environment* (Abingdon: Routledge), pp. 270–86.

Butler, Octavia E. (2005) *Bloodchild and Other Stories* (New York: Seven Stories Press).

Butler, Octavia E. (2019) *Parable of the Sower* (London: Headline). First published in 1993.

Buxbaum, Lara (2017) 'Representations of Xenophobia and Animalisation in *Zebra Crossing, Zoo City* and *Wolf, Wolf*', *Journal of Literary Studies*, 33, 1, pp. 78–93.

Cairns, Rose, Joel Onyango, Andy Stirling, and Phil Johnstone (2022) 'Imagining Urban Transformation in Kenya', *Environmental Science and Policy*, 135, pp. 86–95.

Callamard, Agnès and Kenneth Roth (2022) 'Ethiopia's Invisible Ethnic Cleansing: The World Can't Afford to Ignore Tigray', *Foreign Affairs*, 2 June. Available at https://www.foreignaffairs.com/articles/ethiopia/2022-06-02/ethiopias-invisible-ethnic-cleansing (accessed 9 July 2024).

Caminero-Santangelo, Byron (2014) *Different Shades of Green: African Literature, Environmental Justice, and Political Ecology* (Charlottesville: University of Virginia Press).

Caminero-Santangelo, Byron and Garth Myers (2011) 'Introduction', in Byron Caminero-Santangelo and Garth Myers (eds), *Environment at the Margins: Literary and Environmental Studies in Africa* (Athens, OH: Ohio University Press), pp. 1–21.

Carpenter, Charli (2016) 'Rethinking the Political/-Science-/Fiction Nexus: Global Policy Making and the Campaign to Stop Killer Robots', *Perspectives on Politics*, 14, 1, pp. 53–69.

Carrington, Damian and Èlia Borràs (2024) '"When It's This Hot, Time Stands Still": Surviving West Africa's Blistering Heat', *The Guardian* (UK), 22 March.

Carson, Rachel (1964) *Silent Spring* (London; Hamish Hamilton).

Cassegård, Carl (2024) 'Activism without Hope? Four Varieties of Postapocalyptic Environmentalism', *Environmental Politics*, 33, 3, pp. 444–64.

Castoriadis, Cornelius (1987) *The Imaginary Institution of Society* (Cambridge: Polity).

Chakrabarty, Dipesh (2000) *Provincializing Europe: Postcolonial Thought and Historical Difference* (Princeton: Princeton University Press).

Chakrabarty, Dipesh (2012) 'Postcolonial Studies and the Challenge of Climate Change', *New Literary History*, 43, 1, pp. 1–18.

Chigumadzi, Panashe (2023) 'With What Are You Apologizing?', *Africa Is a Country*, 25 October. Available at https://africasacountry.com/2023/10/with-what-are-you-apologizing (accessed 28 October 2023).

Chimhowu, Admos O., David Hulme, and Lauchlan T. Munro (2019) 'The "New" National Development Planning and Global Development Goals: Processes and Partnerships', *World Development*, 120, pp. 76–89.

Chinigò, Davide (2022) *Everyday Practices of State Building in Ethiopia: Power, Scale, Performativity* (Oxford: Oxford University Press).

Clark, Pilita (2023) 'How to Rethink Climate Change', *Financial Times* [UK], 31 May. Available at https://www.ft.com/content/eef80bc1-c09f-437f-84c7-6feb7a631ba9 (accessed 18 November 2023).

Clark, Timothy (2010) 'Some Climate Change Ironies: Deconstruction, Environmental Politics and the Closure of Ecocriticism', *The Oxford Literary Review*, 32, 1, pp. 131–49.

Clark, Timothy (2015) *Ecocriticism on the Edge: The Anthropocene as a Threshold Concept* (London: Bloomsbury).

Clarke, Jim (2013) 'Reading Climate Change in J. G. Ballard', *Critical Survey*, 25, 2, pp. 7–21.

Cloete, Elsie (2015) 'Johannesburg Zoologica: Reading the Afropolis through the Eyes of Lauren Beukes' *Zoo City*', *Cross/Cultures*, 182, pp. 351–70.

Cole, Matthew Benjamin (2022) '"At the Heart of Human Politics": Agency and Responsibility in the Contemporary Climate Novel', *Environmental Politics*, 31, 1, pp. 132–51.

Constantinou, Costas M. and Eleni Christodoulou (2024) 'On Making Peace with Nature: Visions and Challenges towards an Ecological Diplomacy', *Review of International Studies*, 50, pp. 579–99.

Coogler, Ryan (2018) *Black Panther* [film] (Marvel Studios).

Coogler, Ryan (2022) *Black Panther: Wakanda Forever* [film] (Marvel Studios).

Côté-Roy, Laurence and Sarah Moser (2019) '"Does Africa Not Deserve Shiny New Cities?" The Power of Seductive Rhetoric around New Cities in Africa', *Urban Studies*, 56, 12, pp. 2391–407.

Coward, Martin (2009) *Urbicide: The Politics of Urban Destruction* (London: Routledge).

Crawford, Neta C. (2003) 'Feminist Futures: Science Fiction, Utopia, and the Art of Possibilities in World Politics', in Jutta Weldes (ed.), *To Seek out New Worlds: Exploring Links between Science Fiction and Politics* (New York: Palgrave Macmillan), pp. 195–220.

Cudworth, Erika and Stephen Hobden (2011) *Posthuman International Relations: Complexity, Ecologism and Global Politics* (London: Zed).

Cudworth, Erika and Stephen Hobden (2022) 'Zoonotic Politics: The Impossible Bordering of the Leaky Boundaries of Species', *Millennium*, 50, 3, pp. 647–68.

Dabiri, Emma (2020) *Don't Touch My Hair* (London: Allen Lane).

Daniel III, J. Furman and Paul Musgrave (2017) 'Synthetic Experiences: How Popular Culture Matters for Images of International Relations', *International Studies Quarterly*, 61, 2017, pp. 503–16.

Dauphinee, Elizabeth (2010) 'The Ethics of Autoethnography', *Review of International Studies*, 36, 2010, pp. 799–818.

Davis, Heather and Zoe Todd (2017) 'On the Importance of a Date, or Decolonizing the Anthropocene', *ACME: An International Journal for Critical Geographies*, 16, 4, pp. 761–80.

Davis, Janae, Alex A. Moulton, Levi Van Sant, and Brian Williams (2019) 'Anthropocene, Capitalocene, Plantationocene?: A Manifesto for Ecological Justice in an Age of Global Crises', *Geography Compass*, 13:e12438, pp. 1–15.

Davis, Mike (2006) *Planet of Slums* (London: Verso).

Davis, Mike (2010) 'Who Will Build the Ark?' *New Left Review* II, 61, pp. 29–46.

Death, Carl (2011) 'Leading by Example: South African Foreign Policy and Global Environmental Politics', *International Relations*, 25, 4, pp. 453–76.

Death, Carl (2014) 'Introduction: Critical Environmental Politics', in Carl Death (ed.), *Critical Environmental Politics* (Abingdon: Routledge), pp. 1–12.

Death, Carl (2015) 'Four Discourses of the Green Economy in the Global South', *Third World Quarterly*, 36, 12, pp. 2207–24.

Death, Carl (2016) *The Green State in Africa* (New Haven: Yale University Press).

Death, Carl (2022a) 'Africanfuturist Socio-Climatic Imaginaries and Nnedi Okorafor's Wild Necropolitics', *Antipode*, 54, 1, pp. 240–58.

Death, Carl (2022b) 'Climate Fiction, Climate Theory: Decolonising Imaginations of Global Futures', *Millennium*, 50, 2, pp. 430–55.

Death, Carl (2024a) 'Narrating Transitions to Low Carbon Futures: The Role of Long-term Strategies (LTS) In Fossil Fuel Producing Emerging Economies', *New Political Economy*, 29, 4, pp. 628–45.

Death, Carl (2024b) 'Unfamiliar Families and Disturbing Climate Futures', *International Feminist Journal of Politics*, 26, 2, pp. 264–85.

Delany, Samuel R. (1976) *Triton: An Ambiguous Heterotopia* (New York: Bantam Books).

Delany, Samuel R. (2004) *Stars in My Pocket Like Grains of Sand* (Middletown: Wesleyan University Press).

Delany, Samuel R. (2009) *The Jewel-Hinged Jaw: Notes on the Language of Science Fiction* (Middletown: Wesleyan University Press).

Dery, Mark (1994) 'Black to the Future: Interviews with Samuel R. Delany, Greg Tate, and Tricia Rose', in Mark Dery (ed.), *Flame Wars: The Discourse of Cyberculture* (Durham; Duke University Press), pp. 179–222.

Diagne, Souleymane Bachir (2002) 'Keeping Africanity Open', *Public Culture*, 14, 3, pp. 621–3.

Diene, Mame Bougouma (2022) 'Bittersweet are the waters', in Francesco Verso (ed.) Meteotopia: Futures of Climate (In)Justice (Roma: Associazione Future Fiction), pp. 81–110.

Dowdall, Lisa (2013) 'The Utopian Fantastic in Nnedi Okorafor's Who Fears Death', *Paradoxa*, 25, pp. 173–90.

Duffy, Rosaleen (2010) *Nature Crime: How We Are Getting Conservation Wrong* (New Haven: Yale University Press).

Duncan, Carol B. (2005) 'Hard Labour Religion, Sexuality and the Pregnant Body in the African Diaspora', *Journal for the Association of Research on Mothering*, 7, 1, pp. 167–73.

Dzvimbo, Munyaradzi A., Abraham R. Matamanda, Samuel Adelabu, Adriaan Van der Walt, and Albert Mawonde (2023) 'Understanding the Politics of Climate Change in Zimbabwe', in Walter Leal Filho, Marina Kovaleva, Fátima Alves, and Ismaila Rimi Abubakar (eds), *Climate Change Strategies: Handling the Challenges of Adapting to a Changing Climate* (Cham: Springer), pp. 183–98.

Eatough, Matthew (2015) 'Planning the Future: Scenario Planning, Infrastructural Time, and South African Fiction', *MFS Modern Fiction Studies*, 61, 4, pp. 690–714.

Eatough, Matthew (2017) 'African Science Fiction and the Planning Imagination', *The Cambridge Journal of Postcolonial Literary Inquiry*, 4, 2, pp. 237–57.

Edelman, Lee (2004) *No Future: Queer Theory and the Death Drive* (Durham: Duke University Press).

Edkins, Jenny (2013) 'Novel Writing in International Relations: Openings for a Creative Practice', *Security Dialogue*, 44, 4, pp. 281–97.

Elmandjra, Mahdi (1984) 'Reclaiming the Future: Futures Studies in Africa', *Futures*, 16, 6, pp. 574–8.

Elum, Zelda Anne and Vuyo Mjimba (2020) 'Potential and Challenges of Renewable Energy Development in Promoting a Green Economy in Nigeria', *Africa Review*, 12, 2, pp. 172–91.

Embleton, Stephen (2017) 'Land of Light', *The Manchester Review*, 17 (July). Available at https://www.themanchesterreview.co.uk/?p=7680 (accessed 22 September 2023).

Emmerich, Roland (2004) *The Day After Tomorrow* [film] (20th Century Fox).

Emmerich, Roland (2009) *2012* [film] (Columbia Pictures).

Ericson, Suzanne (2018) 'Thinking with Crocodiles, Thinking through Humans: Vulnerable, Entangled Selves in Lauren Beukes's *Zoo City*', *Scrutiny2*, 23, 1, pp. 22–34.

Eschrich, Joey and Clark A. Miller (eds) (2018) *The Weight of Light: A Collection of Solar Futures* (Phoenix: Arizona State University).

Eshun, Kodwo (2003) 'Further Considerations on Afrofuturism', *CR: The New Centennial Review*, 3, 2, pp. 287–302.

European Commission (2018) *A Clean Planet for All: A European Strategic Long-term Vision for a Prosperous, Modern, Competitive and Climate Neutral Economy*, 28 November.

Evans, Claire L. (2015) 'Climate Change Is So Dire We Need a New Kind of Science Fiction to Make Sense of It', *The Guardian* [UK], 20 August. Available at https://www.theguardian.com/commentisfree/2015/aug/20/climate-change-science-fiction (accessed 5 May 2020).

Evans, Rebecca (2017) 'Fantastic Futures? Cli-fi, Climate Justice, and Queer Futurity', *Resilience: A Journal of the Environmental Humanities*, 4, 2–3, pp. 94–110.

Fakier, Khayaat and Jacklyn Cock (2018) 'Eco-feminist Organizing in South Africa: Reflections on the Feminist Table', *Capitalism Nature Socialism*, 29, 1, pp. 40–57.

Fanon, Frantz (2001) *The Wretched of the Earth* (tr. C. Farrington) (London: Penguin).

Fanon, Frantz (2008) *Black Skin, White Masks* (tr. C. L. Markmann) (London: Pluto).

Federal Democratic Republic of Ethiopia (2023) *Ethiopia's Long-Term Low Emission and Climate Resilient Development Strategy* (2020–2050), 21 June.

Federal Government of Nigeria (2021) *Nigeria's First Nationally Determined Contribution—2021 Update* (Abuja: Federal Ministry of the Environment).

Federal Ministry of Environment, Nigeria (2021) *2050 Long-Term Vision for Nigeria (LTV-2050): Towards the Development of Nigeria's Long-Term Low Emissions Development Strategy (LT-LEDS)*, Department of Climate Change, November. Available at https://unfccc.int/sites/default/files/resource/Nigeria_LTS1.pdf (accessed 24 June 2022).

Feola, Giuseppe (2020) 'Capitalism in Sustainability Transitions Research: Time for a Critical Turn?', *Environmental Innovation and Societal Transitions*, 35, pp. 241–50.

Foucault, Michel and Jay Miskowiec (1986) 'Of Other Spaces', *Diacritics*, 16, 1, pp. 22–7.

Fougner, Tore (2023) 'Animals and Diplomacy: On the Prospect for Interspecies Diplomacy', *International Relations*, 37, 3, pp. 449–74.

Fishel, Stefanie R. (2023) 'The Global Tree: Forests and the Possibility of a Multispecies IR', *Review of International Studies*, 49, 2, pp. 223–40.

Fiskio, Janet (2012) 'Apocalypse and Ecotopia: Narratives in Global Climate Change Discourse', *Race, Gender & Class*, 19, 1/2, pp. 12–36.

Franz, Benedikt (2022) 'Narrative Time and International Relations', *Journal of International Relations and Development*, 25, 2022, pp. 761–83.

Frazier, Chelsea M. (2016) 'Troubling Ecology: Wangechi Mutu, Octavia Butler, and Black Feminist Interventions in Environmentalism', *Critical Ethnic Studies*, 2, 1, pp. 40–72.

Freedman, Carl (2000) *Critical Theory and Science Fiction* (Middleton, CT: Wesleyan University Press).

Gaard, Greta (2004) 'Towards a queer ecofeminism', in Rachel Stein (ed.) *New Perspectives on Environmental Justice: Gender, Sexuality, and Activism* (New Brunswick: Rutgers University Press), pp. 21–44.

Gaard, Greta (2017) *Critical Ecofeminism* (Lanham: Lexington Books).

Gabay, Clive (2018) *Imagining Africa: Whiteness and the Western Gaze* (Cambridge: Cambridge University Press).

Gan, Elaine, Anna Tsing, Heather Swanson, and Nils Bubandt (2017) 'Introduction: Haunted landscapes of the Anthropocene', in Anna Tsing, Heather Swanson, Elaine Gan, and Nils Bubandt (eds), *Arts of Living on a Damaged Planet: Ghosts and Monsters of the Anthropocene* (Minneapolis: University of Minnesota Press), pp. 1–14.

Gani, Jasmine K. and Rabea M. Khan (2024) 'Positionality Statements as a Function of Coloniality: Interrogating Reflexive Methodologies', *International Studies Quarterly*, 68, pp. 1–13.

Gardner, Peter, Olalekan Adekola, Tiago Carvalho, and Thomas O'Brien (2023) 'Confronting the Climate Crisis in Africa: Just Transitions and Extinction Rebellion in Nigeria and South Africa', *Review of African Political Economy*, 50, 177–8, pp. 475–90.

Gardner, Peter, Thomas O'Brien, Tiago Carvalho, and Olalekan Adekola (2024) 'Extinction Rebellion in the Gambia', *Social Movement Studies*, 23, 1, pp. 122–8.

Gee, Maggie (1998) *The Ice People* (London: Metro).

Geels, Frank W. (2014) 'Regime Resistance against Low-Carbon Transitions: Introducing Politics and Power into the Multi-Level Perspective', *Theory, Culture & Society*, 31, 5, pp. 21–40.

Gergan, Mabel, Sara Smith, and Pavithra Vasudevan (2020) 'Earth beyond Repair: Race and Apocalypse in Collective Imagination', *EPD: Society and Space*, 38, 1, pp. 91–110.

Ghosh, Amitav (2016) *The Great Derangement: Climate Change and the Unthinkable* (Chicago: University of Chicago Press).

Ghosh, Amitav (2021) *The Nutmeg's Curse: Parables for a Planet in Crisis* (London: John Murray).

Gibson, Julia D. (2020) 'Stories We Tell about the End of the World: (Post)Apocalyptic Climate Fiction Working towards Climate Justice', *Paradoxa*, 31, pp. 205–29.

Gilio-Whitaker, Dina (2019) *As Long as Grass Grows: The Indigenous Fight for Environmental Justice, from Colonization to Standing Rock* (Boston: Beacon Press).

Gilroy, Paul (2002) 'Toward a Critique of Consumer Imperialism', *Public Culture*, 14, 3, pp. 589–91.

Gilroy, Paul (2005) *Postcolonial Melancholia* (New York: Columbia University Press).

Goodbody, Axel and Adeline Johns-Putra (2019) 'The Rise of the Climate Change Novel', in Adeline Johns-Putra (ed.), *Climate and Literature* (Cambridge: Cambridge University), pp. 229–45.

Gore, Ellie (2018) 'Reflexivity and Queer Embodiment: Some Reflections on Sexualities Research in Ghana', *Feminist Review*, 120, 1, pp. 101–19.

Government of The Gambia (2022) *The Gambia's Long-Term Climate-Neutral Development Strategy 2050* (Banjul: Ministry of Environment, Climate Change and Natural Resources).

Government of Zimbabwe (2022) *Zimbabwe Long-term Low Greenhouse Gas Emission Development Strategy (2020–2050)* (Harare: Climate Change Management Department).

Granjou, Céline (2016) *Environmental Changes: The Futures of Nature* (London: ISTE Press).

Granjou, Céline and Juan Francisco Salazar (2016) 'Future', *Environmental Humanities*, 8, 2, pp. 240–4.

Grasso, Marco (2024) 'The Case for Climate Reparations by Fossil Fuel Companies: Ethical Foundations, Monetary Estimates and Feasibility', *Development and Change*, 55, 4, pp. 727–51.

Grayson, Kyle, Matt Davies, and Simon Philpott (2009) 'Pop Goes IR? Researching the Popular Culture–World Politics Continuum', *Politics*, 29, 3, pp. 155–63.

Green, Jessica F. and Raúl Salas Reyes (2023) 'The History of Net Zero: Can We Move from Concepts to Practice?', *Climate Policy*, 23, 7, pp. 901–15.

Green-Simms, Lindsey (2016) 'The Emergent Queer: Homosexuality and Nigerian Fiction in the 21st Century', *Research in African Literatures*, 47, 2, pp. 139–61.

Grosz, Elizabeth (2004) *In the Nick of Time: Politics, Evolution and the Untimely* (Durham: Duke University Press).

Haiven, Max (2014) *Crises of Imagination, Crises of Power: Capitalism, Creativity and the Commons* (London: Zed Books).

Haiven, Max and Alex Khasnabish (2014) *The Radical Imagination: Social Movement Research in the Age of Austerity* (London: Zed Books).

Hajer, Maarten and Wytske Versteeg (2019) 'Imagining the Post-fossil City: Why Is It So Difficult to Think of New Possible Worlds?', *Territory, Politics, Governance*, 7, 2, pp. 122–34.

Hall, Melinda (2019) 'What Future People Will There Be? Neurodiverse Heroes for a Changing Planet', *Journal of Science Fiction*, 3, 2, pp. 15–17.

Hamann, M., R. Biggs, L. Pereira, R. Preiser, T. Hichert, R. Blanchard, H. Warrington-Coetzee, N. King, A. Merrie, W. Nilsson, P. Odendaal, S. Poskitt, D. Sanchez Betancourt, and G. Ziervogel (2020) 'Scenarios of Good Anthropocenes in Southern Africa', *Futures*, 118, 102526, pp. 1–16.

Hammond, Marit (2021) 'Imagination and Critique in Environmental Politics', *Environmental Politics*, 30, 1–2, pp. 285–305.

Hanson, Clare (2016) 'A Catastrophic Universe: Lessing, Posthumanism and Deep History', in Kevin Brazil, David Sergeant, and Tom Sperlinger (eds), *Doris Lessing and the Forming of History* (Edinburgh: Edinburgh University Press), pp. 164–80.

Haraway, Donna J. (2016) *Staying with the Trouble: Making Kin in the Chthulucene* (Durham: Duke University Press).

Harper, Tyler (2023) 'What *The Last of Us*, *Snowpiercer* and "Climate Fiction" Get Wrong', BBC (UK), 18 April. Available at https://www.bbc.com/culture/article/20230418-what-snowpiercer-and-climate-fiction-get-wrong (accessed 18 November 2023).

Harris, Dylan M. (2017) 'Telling the Story of Climate Change: Geologic Imagination, Praxis, and Policy', *Energy Research & Social Science*, 31, pp. 179–83.

Hartmann, Ivor W. (2015) 'Last wave', *Jalada*, 2. Available at https://jaladaafrica.org/2015/01/15/last-wave-by-ivor-w-hartmann/ (accessed 22 September 2023).

Hatzisavvidou, Sophia (2024) 'Envisioning Ecopolitical Futures: Reading Climate Fiction as Political Theory', *Futures*, 163, pp. 1–34.

Heise, Ursula K. (2008) *Sense of Place and Sense of Planet: The Environmental Imagination of the Global* (Oxford: Oxford University Press).

Helford, Elyce Rae (1994) '"Would You Really Rather Die than Bear My Young?": The Construction of Gender, Race, and Species in Octavia E. Butler's "Bloodchild"', *African American Review*, 28, 2, pp. 259–71.

Helliker, Kirk and Tendai Murisa (2020) 'Zimbabwe: Continuities and Changes', *Journal of Contemporary African Studies*, 38, 1, pp. 5–17.

Heron, Kai (2024) 'Capitalist Catastrophism and Eco-apartheid', *Geoforum*, 153, 103874, pp. 1–10.

Hetherington, Kevin (1997) *The Badlands of Modernity: Heterotopia and Social Ordering* (London: Routledge).

Hetherington, Kevin (2011) 'Foucault, the Museum and the Diagram', *The Sociological Review*, 59, 3, pp. 457–75.

Hindess, Barry (2007) 'The Past Is Another Culture', *International Political Sociology*, 1, pp. 325–38.

Hintjens, Helen M. (1999) 'Explaining the 1994 Genocide in Rwanda', *Journal of Modern African Studies*, 37, 2, pp. 241–86.

Hoad, Neville (2007) *African Intimacies: Race, Homosexuality and Globalization* (Minneapolis: University of Minnesota Press).

Hoad, Neville (2016) 'Dystopian Futures, Apartheid, and Postapartheid Allegories: Contemporary Imaginings of Johannesburg', *Comparative Studies of South Asia, Africa and the Middle East*, 36, 2, pp. 293–306.

Hochstetler, Kathryn (2020) *Political Economies of Energy Transition: Wind and Solar Power in Brazil and South Africa* (Cambridge: Cambridge University Press).

Hoelscher, Kristian, Nick Dorward, Sean Fox, Taibat Lawanson, Jeffrey W. Paller, and Melanie L. Phillips (2023) 'Urbanization and Political Change in Africa', *African Affairs*, 122, 488, pp. 353–76.

Hom, Andrew R. (2020) *International Relations and the Problem of Time* (Oxford: Oxford University Press).

Hom, Andrew R. and Brent J. Steele (2016) 'Child's Play: Temporal Discourse, Counterpower, and Environmental Politics', in Anna M. Agathangelou and Kyle D. Killian (eds), *Time, Temporality and Violence in International Relations: (De)Fatalizing the Present, Forging Radical Alternatives* (Abingdon: Routledge), pp. 189–204.

Hopkinson, Nalo (2004) 'Introduction', in Nalo Hopkinson and Uppinder Mehan (eds), *So Long Been Dreaming: Postcolonial Science Fiction & Fantasy* (Vancouver: Arsenal Pulp Press), pp. 7–9.

Horkheimer, Max and Theodor W. Adorno (2002) *Dialectics of Enlightenment: Philosophical Fragments* (Stanford: Stanford University Press).

Howell, Jessica (2019) 'Climate and Race in the Age of Empire', in Adeline Johns-Putra (ed.), *Climate and Literature* (Cambridge: Cambridge University), pp. 163–78.

Hoydis, Julia (2017) 'A Darker Shade of Justice: Violence, Liberation, and Afrofuturist Fantasy in Nnedi Okorafor's *Who Fears Death*', in Anke Bartels, Lars Eckstein, Nicole Waller, and Dirk Wiemann (eds), *Postcolonial Justice* (Leiden: Brill), pp. 179–99.

Huggan, Graham and Helen Tiffin (2007) 'Green Postcolonialism', *Interventions: International Journal of Postcolonial Studies*, 9, 1, pp. 1–11.

Hugo, Esthie (2017) 'Looking Forward, Looking Back: Animating Magic, Modernity and the African city-future in Nnedi Okorafor's *Lagoon*', *Social Dynamics*, 43, 1, pp. 46–58.

Hulme, Mike (2010) 'Cosmopolitan Climates: Hybridity, Foresight and Meaning', *Theory, Culture & Society*, 27, 2–3, pp. 267–76.

Hultin, Niklas and Tone Sommerfelt (2020) 'Anticipatory Tribalism: Accusatory Politics in the "New Gambia"', *Journal of Modern African Studies*, 58, 2, pp. 257–79.

Hunt, Eileen M. (2023) 'Mary Shelley's *The Last Man*: Existentialism and IR Meet the Post-apocalyptic Pandemic Novel', *Review of International Studies*, 49, 5, pp. 832–54.

Hunter, Megan (2017) *The End We Start From* (London: Picador).

Hutchings, Kimberly (2008) *Time and World Politics: Thinking the Present* (Manchester: Manchester University Press).

Iheka, Cajetan (2017) *Naturalizing Africa: Ecological Violence, Agency, and Postcolonial Resistance in African Literature* (Cambridge: Cambridge University Press).

Iheka, Cajetan (2019) 'African Literature, Audience and the Search for the (Non)human', in Moradewun Adejunmobi and Carli Coetzee (eds), *Routledge Handbook of African Literature* (Abingdon: Routledge), pp. 203–14.

Iheka, Cajetan (2021) *African Ecomedia: Network Forms, Planetary Politics* (Durham: Duke University Press).

Illich, Ivan (1973) *Tools for Conviviality* (London: Marion Boyars).

Ize-Iyamu, Osahon (2019) 'More Sea than Tar', *The Reckoning*, 28 February. Available at https://reckoning.press/more-sea-than-tar/ (accessed 3 August 2023).

Imarisha, Walidah and Adrienne Maree Brown (eds) (2015) *Octavia's Brood: Science Fiction Stories from Social Justice Movements* (Oakland, CA: AK Press).

Inayatullah, Naeem and Elizabeth Dauphinee (2016) 'Permitted Urgency: A Prologue', in Naeem Inayatullah and Elizabeth Dauphinee (eds), *Narrative Global Politics: Theory, History and the Personal in International Relations* (London: Routledge), pp. 1–4.

Jackson, Zakiyyah Iman (2020) *Becoming Human: Matter and Meaning in an Antiblack World* (New York: NYU Press).

Jackson, Sandra and Julie Moody-Freeman (2009) 'The Genre of Science Fiction and the Black Imagination', *African Identities*, 7, 2, pp. 127–32.

Jameson, Frederic (2007) *Archaeologies of the Future: The Desire Called Utopia and Other Science Fictions* (London: Verso).

Jasanoff, Sheila (2010) 'A New Climate for Society', *Theory, Culture and Society*, 27, 2–3, pp. 233–53.

Jasanoff, Sheila (2015) 'Future Imperfect: Science, Technology, and the Imaginations of Modernity', in Sheila Jasanoff and Sang-Hyun Kim (eds), *Dreamscapes of Modernity: Sociotechnical Imaginaries and the Fabrication of Power* (Chicago: University of Chicago Press), pp. 1–33.

Jessop, Bob (2010) 'Cultural Political Economy and Critical Policy Studies', *Critical Policy Studies*, 3, 3–4, pp. 336–56.

Johnson, Peter (2006) 'Unravelling Foucault's "Different Spaces"', *History of the Human Sciences*, 19, 4, pp. 75–90.

Johns-Putra, Adeline (2015) '"Cli-fi" Novels Humanise the Science of Climate Change—and Leading Authors Are Getting in on the Act', *The Conversation*, 26 November. Available at https://theconversation.com/cli-fi-novels-humanise-the-science-of-climate-change-and-leading-authors-are-getting-in-on-the-act-51270 (accessed 17 November 2023).

Johns-Putra, Adeline (2016) 'Climate Change in Literature and Literary Studies: From Cli-fi, Climate Change Theater and Ecopoetry to Ecocriticism and Climate Change Criticism', *WIREs Climate Change*, 7, 2, pp. 266–82.

Johns-Putra, Adeline (2018) 'The Rest Is Silence: Postmodern and Postcolonial Possibilities in Climate Change Fiction', *Studies in the Novel*, 50, 1, pp. 26–42.

Johns-Putra, Adeline (2019) *Climate Change and the Contemporary Novel* (Cambridge: Cambridge University Press).

Jones, Calvert W. and Celia Paris (2018) 'It's the End of the World and They Know It: How Dystopian Fiction Shapes Political Attitudes', *Perspectives on Politics*, 16, 4, pp. 969–89.

Joo, Hee-Jung S. (2020) 'We Are the World (but Only at the End of the World): Race, Disaster, and the Anthropocene', *EPD: Society and Space*, 38, 1, pp. 72–90.

Jue, Melody (2017) 'Intimate Objectivity: On Nnedi Okorafor's Oceanic Afrofuturism', *Women's Studies Quarterly*, 45, 1/2, pp. 171–88.

Julien, Eileen (2006) 'The Extroverted African Novel', in Franco Moretti (ed.), *The Novel, Volume 1: History, Geography, and Culture* (Princeton: Princeton University Press), pp. 667–700.

Kafer, Alison (2013) *Feminist, Queer, Crip* (Bloomington: Indiana University Press).

Kahiu, Wanuri (2009) *Pumzi* [film] (Inspired Minority Pictures).

Kaplan, E. Ann (2016) *Climate Trauma: Foreseeing the Future in Dystopian Film and Fiction* (New Jersey: Rutgers University Press).

Kapstein, Helen (2016) 'Crude Fictions: How New Nigerian Short Stories Sabotage Big Oil's Master Narrative', *Postcolonial Text*, 11, 1, pp. 1–18.

Karam, Beschara and Mark Kirby-Hirst (2019) 'Guest Editorial for Themed Section *Black Panther* and Afrofuturism: Theoretical Discourse and Review', *Image and Text*, 33, pp. 1–15.

Kelley, Robin D. G. (2002) *Freedom Dreams: The Black Radical Imagination* (Boston: Beacon Press).

Kermode, Frank (2000) *The Sense of an Ending: Studies in the Theory of Fiction* (Oxford: Oxford University Press).

Kaboub, Fadhel (2023) 'Rich Countries Are in Debt Default: COP28 Traps and Blind Spots', *Global South Perspectives*, 12 December. Available at https://globalsouthperspectives.substack.com/p/rich-countries-are-in-debt-default (accessed 24 January 2024).

Kanitkar, Tejal, Akhil Mythri, and T. Jayaraman (2024) 'Equity Assessment of Global Mitigation Pathways in the IPCC Sixth Assessment Report', *Climate Policy*, 24, 8, pp. 1129–48.

Keita, Lamin (2023) 'Youth and Protest: How "#Gambia" Ended Decades of Autocratic Rule', *Canadian Journal of African Studies/Revue canadienne des études africaines*, 57, 3, pp. 585–604.

Kiersey, Nicholas J. and Iver B. Neumann (2015) 'Worlds of Our Making in Science Fiction and International Relations', in Federica Caso and Caitlin Hamilton (eds), *Popular Culture and World Politics: Theories, Methods, Pedagogies* (Bristol: E-IR), pp. 74–82.

Klein, Naomi (2019) 'A Message from the Future with Alexandria Ocasio-Cortez', *The Intercept*, 17 April. Available at https://theintercept.com/2019/04/17/green-new-deal-short-film-alexandria-ocasio-cortez/ (accessed 5 November 2023).

Kotecki, Kristine (2020) 'Apocalyptic Affect in Nnedi Okorafor's Speculative Futures', *Research in African Literatures*, 51, 3, pp. 164–79.

Krishnan, Madhu (2014) *Contemporary African Literature in English: Global Locations, Postcolonial Identifications* (Basingstoke: Palgrave Macmillan).

Kristeva, Julia (1986) 'Stabat Mater', in Moi, Toril (ed.) *The Kristeva Reader* (New York: CUP), pp. 160–86.

Kuznetski, Julia (2021) 'Ecofeminist (Post) Ice-Age Ecotopia: Doris Lessing's Mara and Dann Books', in Douglas A. Vakoch (ed.), *Dystopias and Utopias on Earth and Beyond: Feminist Ecocriticism of Science Fiction* (New York: Routledge), pp. 50–66.

Lachapelle, Erick, Robert MacNeil, and Matthew Paterson (2017) 'The Political Economy of Decarbonisation: from Green Energy "Race" to Green "Division of Labour"', *New Political Economy*, 22, 3, pp. 311–27.

Laduke, Winona (2004) 'Foreword', in Rachel Stein (ed.), *New Perspectives on Environmental Justice: Gender, Sexuality, and Activism* (New Brunswick: Rutgers University Press), p. xiii.

Langer, Jessica (2011) *Science Fiction and Postcolonialism* (New York: Palgrave Macmillan).

Lavender III, Isiah (2011) *Race in American Science Fiction* (Bloomington: Indiana University Press).

Lavers, Tom (2024) *Ethiopia's 'Developmental State': Political Order and Distributive Crisis* (Cambridge: Cambridge University Press).

Le Guin, Ursula K. (2016) 'A War without End', in Thomas More, Ursula K. Le Guin, and China Miéville, *Utopia* (London: Verso), pp. 356–77.

Lessing, Doris (1999) *Mara and Dann: An Adventure* (London: Flamingo).

Lessing, Doris (2006) *The Story of General Dann and Mara's Daughter, Griot and the Snow Dog* (London: Harper Perennial).

Lessing, Doris (2007) *Nobel Prize speech*, 7 December. Available at https://www.nobelprize.org/prizes/literature/2007/lessing/25434-doris-lessing-nobel-lecture-2007/ (accessed 6 July 2023).

Lessing, Doris (2014) *The Golden Notebook* (London: Fourth Estate). First published 1962.

Lessing, Doris (2017) *Web of Stories—Life Stories of Remarkable People* [video], 8 May. Available at https://www.youtube.com/watch?v=hUdDjaQzH9c&list=PLVV0r6CmEsFxnVS80V-yoG3_GTTZdvdw7 (accessed 28 June 2023).

Levitas, Ruth (2013) *Utopia as Method: The Imaginary Reconstitution of Society* (New York: Palgrave Macmillan).

Levy, David L. and André Spicer (2013) 'Contested Imaginaries and the Cultural Political Economy of Climate Change', *Organization*, 20, 5, pp. 659–78.

Lewis, Sophie (2019) *Full Surrogacy Now: Feminism against Family* (London: Verso).

Liboiron, Max (2021) *Pollution Is Capitalism* (Durham: Duke University Press).

Ling, L. H. M. (2014) *Imagining World Politics: Sihar and Shenya, a Fable for Our Times* (Abingdon: Routledge).

Liveley, Genevieve (2022) 'Narrative: Telling Social Futures', in Carlos López Galviz and Emily Spiers (eds), *Routledge Handbook of Social Futures* (London: Routledge), pp. 224–32.

Llansó, Miguel (2015) *Crumbs [film]* (Lanzadera and Birabiro Films).

Luke, Timothy W. (2015) 'The Climate Change Imaginary', *Current Sociology Monograph*, 63, 2, pp. 280–96.

Lundborg, Tom (2016) 'Time', in Aoileann Ní Mhurchú and Reiko Shindo (eds), *Critical Imaginations in International Relations* (Abingdon: Routledge), pp. 262–76.

Luxton, Hopkinson Maathai, Wangari Muta (2007) *Unbowed: A Memoir* (London: William Heinemann).

Maathai, Wangari Muta (2009) *The Challenge for Africa* (New York: Random House).

Macdonald, Graeme (2017) '"Monstrous Transformer": Petrofiction and World Literature', *Journal of Postcolonial Writing*, 53, 3, pp. 289–302.

MacGregor, Sherilyn (2006) *Beyond Mothering Earth: Ecological Citizenship and the Politics of Care* (Vancouver: UBC Press).

MacGregor, Sherilyn (2010) 'Gender and Climate Change: From Impacts to Discourses', *Journal of the Indian Ocean Region*, 6, 2, pp. 223–38.

MacGregor, Sherilyn (2014) 'Only Resist: Feminist Ecological Citizenship and the Post-Politics of Climate Change', *Hypatia*, 29, 3, pp. 617–33.

Machin, Amanda (2022) 'Climates of Democracy: Skeptical, Rational, and Radical Imaginaries', *WIREs Climate Change*, e774, pp. 1–13.

Mackay, Alistair (2022) *It Doesn't Have to Be This Way* (Cape Town: Kwela Books).

Mackey, Allison (2018) 'Guilty Speculations: The Affective Climate of Global Anthropocene Fictions', *Science Fiction Studies*, 45, 3, pp. 530–44.

Malm, Andreas (2021) *How to Blow Up a Pipeline: Learning to Fight in a World on Fire* (London: Verso).

Mama, Amina (2001) 'Challenging Subjects: Gender and Power in African Contexts', *African Sociological Review/Revue Africaine de Sociologie*, 5, 2, pp. 63–73.

Maniates, Michael (2001) 'Individualization: Plant a Tree, Buy a Bike, Save the World?' *Global Environmental Politics*, 1, 3, pp. 31–52.

Manzo, Kate (2010) 'Imaging Vulnerability: The Iconography of Climate Change', *Area*, 42, 1, pp. 96–107.

Marquardt, Jens and Naghmeh Nasiritousi (2022) 'Imaginary Lock-ins in Climate Change Politics: The Challenge to Envision a Fossil-Free Future', *Environmental Politics*, 31, 4, pp. 621–42.

Matebeni, Zethu and Jabu Pereira (2014) 'Preface', in Zethu Matebeni (eds), *Reclaiming Afrikan: Queer Perspectives on Sexual and Gender Identities* (Athlone: Modjaji Books), pp. 7–9.

Mbembe, Achille (2001) *On the Postcolony* (Berkeley: University of California Press).

Mbembe, Achille (2002) 'African Modes of Self-Writing', translated by Steven Rendall, *Public Culture*, 14, 1, pp. 239–73.

Mbembe, Achille (2004) 'Aesthetics of Superfluity', *Public Culture*, 16, 3, pp. 373–405.

Mbembe, Achille (2006) 'On the Postcolony: A Brief Response to Critics', *African Identities*, 4, 2, pp. 143–78.

Mbembe, Achille (2016) 'Africa in the New Century', in Lien Heidenreich-Seleme and Sean O'Toole (eds), *African Futures: Thinking about the Future through Word and Image* (Bielefield: Kerber Verlag), pp. 315–35.

Mbembe, Achille (2017) *Critique of Black Reason*, translated by Laurent Dubois (Durham: Duke University Press).

Mbembe, Achille (2019) *Necropolitics*, translated by Steven Corcoran (Durham: Duke University Press).

Mcbride, Sindi-Leigh (2023) 'Embodied Climate Knowledge in African Cli-Fi: Alistair Mackay's *It Doesn't Have to Be This Way* (2022) and Nnedi Okorafor's *Noor* (2021)', *ISLE: Interdisciplinary Studies in Literature and Environment*, isad067, pp. 1–22.

McCarthy, Cormac (2006) *The Road* (New York: Alfred A. Knopf).

McClintock, Anne (1995) *Imperial Leather: Race, Gender and Sexuality in the Colonial Contest* (London: Routledge).

McConnell, Bonnie B. and Sheikh Omar Jallow (2022) 'Climate Change Adaptation in The Gambia: The Role of Kanyeleng Communication and Performance', *African Studies Review*, 65, 3, pp. 692–716.

McEwan, Ian (2010) *Solar* (London: Jonathan Cape).

McKibben, Bill (2013) 'Don't Imagine the Future—It's Already Here', *Organization*, 20, 5, pp. 745–7.

McKibben, Bill (2019) 'Hello from the Year 2050. We Avoided the Worst of Climate Change—but Everything Is Different', *Time Magazine* (US), 12 September.

McKittrick, Katherine (2006) *Demonic Grounds: Black Women and the Cartographies of Struggle* (Minneapolis: University of Minnesota Press).

McKittrick, Katherine (ed.) (2015) *Sylvia Wynter: On Being Human as Praxis* (Durham: Duke University Press).

Medhin, Haileselassie and Alemu Mekonnen (2019) 'Green and Climate-Resilient Transformation in Ethiopia', in Fantu Cheru, Christopher Cramer, and Arkebe Oqubay (eds), *The Oxford Handbook of the Ethiopian Economy* (Oxford: Oxford University Press), pp. 288–307.

Mehnert, Antonia (2016) *Climate Change Fictions: Representations of Global Warming in American Literature* (London: Palgrave Macmillan).

Melzer, Patricia (2006) *Alien Constructions: Science Fiction and Feminist Thought* (Austin: University of Texas Press).

Menzies, Ella (2022) 'The World away from the Rain', *Imagine 2200*, 4 October. Available at https://grist.org/fix/climate-fiction/imagine-2200-world-away-from-the-rain/ (accessed 24 January 2024).

Miéville, China (2009) 'Afterword. Cognition as Ideology: A Dialectic of SF theory', in Mark Bould and China Miéville (eds), *Red Planets: Marxism and Science Fiction* (London: Pluto), pp. 231–48.

Miéville, China (2016) 'The Limits of Utopia', in Thomas More, Ursula K. Le Guin, and China Miéville, *Utopia* (London: Verso), pp. 22–52.

Milkoreit, Manjana (2016) 'The Promise of Climate Fiction: Imagination, Story-telling, and the Politics of the Future', in Paul Wapner and Hilal Elver (eds), *Reimagining Climate Change* (Abingdon: Routledge), pp. 171–91.

Milkoreit, Manjana (2017) 'Imaginary Politics: Climate Change and Making the Future', *Elementa: Science of the Anthropocene*, 5, 62, pp. 1–18.

Milkoreit, Manjana (2019) 'Pop-cultural Mobilization: Deploying Game of Thrones to Shift US Climate Change Politics', *International Journal of Politics, Culture and Society*, 32, pp. 61–82.

Miller, Clark A., Jason O'Leary, Elisabeth Graffy, Ellen B. Stechel, and Gary Dirks (2015) 'Narrative Futures and the Governance of Energy Transitions', *Futures*, 70, pp. 65–74.

Millet, Lydia (2020) *A Children's Bible* (New York: W. W. Norton).

Milner, Andrew and J. R. Burgmann (2018) 'Climate Fiction: A World-Systems Approach', *Cultural Sociology*, 12, 1, pp. 22–36.

Mitchell, Audra (2016) 'Beyond Biodiversity and Species: Problematizing Extinction', *Theory, Culture & Society*, 33, 5, pp. 23–42.

Mitchell, Audra (2017) 'Is IR Going Extinct?', *European Journal of International Relations*, 23, 1, pp. 3–25.

Mitchell, Audra (2023) 'Nonhuman, More-Than-Human, and Post-Human International Relations and International Studies', *Oxford Research Encyclopaedia: International Studies*, January. Available at https://doi.org/10.1093/acrefore/9780190846626.013.754 (accessed 18 July 2023).

Mitchell, Audra and Aadita Chaudhury (2020) 'Worlding beyond "the" "End" of "the World": White Apocalyptic Visions and BIPOC Futurisms', *International Relations*, 34, 3, pp. 309–32.

Moezzi, Mithra, Kathryn B. Janda, and Sea Rotmann (2017) 'Using Stories, Narratives, and Storytelling in Energy and Climate Change Research', *Energy Research & Social Science*, 31, pp. 1–10.

Mọgọ, Ebele (2022) 'Omaliyi', *Imagine 2200*, 4 October. Available at https://grist.org/fix/climate-fiction/imagine-2200-omaliyi/ (accessed 24 January 2024).

Mohamed, Najma and Gaylor Montmasson-Clair (2019) 'Policies for Sustainability Transitions in South Africa: A Critical Review', in Najma Mohamed (ed.), *Sustainability Transitions in South Africa* (London: Routledge), pp. 80–100.

Molelekwa, Thabo (2022) 'Unpacking South Africa's Renewable Energy Project', *Mail and Guardian*, 23 September. Available at https://mg.co.za/environment/2022-09-23-unpacking-south-africas-renewable-energy-project/ (accessed 6 October 2022).

Moore, Michele-Lee and Manjana Milkoreit (2020) 'Imagination and Transformations to Sustainable and Just Futures', *Elementa: Science of the Anthropocene*, 8, 1, pp. 1–17.

Moonsamy, Nedine (2016) 'Life Is a Biological Risk: Contagion, Contamination, and Utopia in African Science Fiction', *Cambridge Journal of Postcolonial Literary Inquiry*, 3, 3, pp. 329–43.

Moonsamy, Nedine (2020) 'Fish out of Water: Black Superheroines in Nnedi Okorafor's Lagoon', *Transition*, 129, pp. 175–89.

Moran, Greta (2019) 'The Lion King Missed an Opportunity to Talk about the Climate Crisis', *The Guardian* (UK), 20 August. Available at https://www.theguardian.com/commentisfree/2019/aug/20/lion-king-climate-crisis-scar-pride-lands (accessed 8 March 2024).

More, Thomas (2016) 'Utopia', in Thomas More, Ursula K. Le Guin, and China Miéville, *Utopia* (London: Verso), pp. 54–294. First published in 1516.

Moulin, Carolina (2016) 'Narrative', in Aoileann Ní Mhurchú and Reiko Shindo (eds), *Critical Imaginations in International Relations* (Abingdon: Routledge), pp. 136–52.

Moylan, Tom (2020) 'The Necessity of Hope in Dystopian Times: A Critical Reflection', *Utopian Studies*, 31, 1, pp. 164–93.

Müller, Franziska, Simone Claar, Manuel Neumann, and Carsten Elsner (2020) 'Is Green a Pan-African Colour? Mapping African Renewable Energy Policies and Transitions in 34 Countries', *Energy Research & Social Science*, 68, 101551, pp. 1–9.

Muñoz, José Esteban (2009) *Cruising Utopia: The Then and There of Queer Futurity* (New York: NYU Press).

Munshi, Debashish, Priya A. Kurian, John Foran, and Kum-Kum Bhavnani (2019) 'The Future Is Ours to Seek: Changing the Inevitability of Climate Chaos to Prospects of Hope and Justice', in Kum-Kum Bhavnani, John Foran, Priya A. Kurian, and Debashish Munshi (eds), *Climate Futures: Re-imagining Global Climate Justice* (London: Zed), pp. 1–8.

Musila, Grace A. (2014) 'Laughing at the Rainbow's Cracks? Blackness, Whiteness and the Ambivalences of South African Stand-up Comedy', in Ebenezer Obadare and Wendy Willems (eds), *Civic Agency in Africa: Arts of Resistance in the 21st Century* (Woodbridge: James Currey), pp. 147–66.

Musila, Grace A. (2022) 'Thirteen Ways of Reading African Popular Culture', in Grace A. Musila (ed.), *Routledge Handbook of African Popular Culture* (Abingdon: Routledge), pp. 1–28.

Myers, Garth (2011) *African Cities: Alternative Visions of Urban Theory and Practice* (London: Zed).

Myers, Garth (2018) 'The Future and Africa's Cities', in Tony Binns, Kenneth Lynch, and Etienne Nel (eds), *The Routledge Handbook of African Development* (London: Routledge), pp. 465–73.

National Climate Change Council (2023) *Nigeria's Long-Term Low-Carbon Development Strategy* (Abuja: NCCC).

NDC Partnership (2021) *Nigeria's Updated NDC Is Ambitious and Inclusive*, 9 September. Available at https://ndcpartnership.org/news/nigeria%E2%80%99s-updated-ndc-ambitious-and-inclusive (accessed 5 October 2022).

Ndlovu-Gatsheni, Sabelo J. (2009) 'Making Sense of Mugabeism in Local and Global Politics: "So Blair, Keep Your England and Let Me Keep My Zimbabwe"', *Third World Quarterly*, 30, 6, pp. 1139–58.

Ndlovu-Gatsheni, Sabelo J. (2014) 'Global Coloniality and the Challenges of Creating African Futures', *Strategic Review for Southern Africa*, 36, 2, pp. 181–202.

Net Zero Tracker (2024) *Data Explorer*. Available at https://zerotracker.net/ (accessed 28 October).

Newell, Stephanie (1997) 'Introduction. Writing African Women: Gender, popular culture and literature in West Africa', in Stephanie Newell (ed.), *Writing African Women: Gender, Popular Culture and Literature in West Africa* (London: Zed Books), pp. 1–8.

Newell, Peter (2019) 'Trasformismo or Transformation? The Global Political Economy of Energy Transitions', *Review of International Political Economy*, 26, 1, pp. 25–48.

Newell, Peter (2020) 'Towards a Global Political Economy of Transitions: A Comment on the Transitions Research Agenda', *Environmental Innovation and Societal Transitions*, 34, pp. 344–5.

Newell, Peter (2021) *Power Shift: The Global Political Economy of Energy Transitions* (Cambridge: Cambridge University Press).

Newell, Peter and Harriet Bulkeley (2017) 'Landscape for Change? International Climate Policy and Energy Transitions: Evidence from sub-Saharan Africa', *Climate Policy*, 17, 5, pp. 650–63.

Newell, Peter and Jon Phillips (2016) 'Neoliberal Energy Transitions in the South: Kenyan Experiences', *Geoforum*, 74, pp. 39–48.

Newell, Peter, Matthew, Paterson, and Martin, Craig (2021) 'The Politics of Green Transformations: An Introduction to the Special Section', *New Political Economy*, 26, 6, pp. 903–6.

Nfah-Abbenyi, Juliana Makuchi (1997) *Gender in African Women's Writing: Identity, Sexuality and Difference* (Bloomington: Indiana University Press).

Ngemntu, Vuyokazi (2022) 'The Serpent's Handmaiden', *The Kalahari Review*, 21 December. Available at https://kalaharireview.com/the-serpents-handmaiden-7a4c8b6af7f8 (accessed 3 August 2023).

Ní Mhurchú, Aoileann (2016) 'Thinking with Reproduction: Maternal Time, Citizenship, Migration and Political Subjectivity', *Subjectivity*, 9, pp. 17–37.

Nielsen, Morten, Jason Sumich, and Bjørn Enge Bertelsen (2021) 'Enclaving: Spatial Detachment as an Aesthetics of Imagination in an Urban Sub-Saharan African Context', *Urban Studies*, 58, 5, pp. 881–902.

Nikoleris, Alexandra, Johannes Stripple, and Paul Tenngart (2017) 'Narrating Climate Futures: Shared Socioeconomic Pathways and Literary Fiction', *Climatic Change*, 143, pp. 307–19.

Nikoleris, Alexandra, Johannes Stripple, and Paul Tenngart (2019) 'The "Anthropocene" in Popular Culture: Narrating Human Agency, Force, and Our Place on Earth', in Frank Biermann and Eva Lövbrand (eds), *Anthropocene Encounters: New Directions in Green Political Thinking* (Cambridge: Cambridge University Press), pp. 67–83.

Nixon, Rob (2011) 'Slow Violence, Gender, and the Environmentalism of the Poor', in Byron Caminero-Santangelo and Garth Myers (eds), *Environment at the Margins: Literary and Environmental Studies in Africa* (Athens, OH: Ohio University Press), pp. 257–85.

Nussbaum, Martha (2017) 'Cultivating Humanity: The Narrative Imagination', in Julie Rivkin and Michael Ryan (eds), *Literary Theory: An Anthology* (Chichester: Wiley Blackwell), pp. 481–502. First published in 1997.

Nwozor, Agaptus, Segun Oshewolo, Gbenga Owoeye, and Onjefu Okidu (2021) 'Nigeria's Quest for Alternative Clean Energy Development: A Cobweb of Opportunities, Pitfalls and Multiple Dilemmas', *Energy Policy*, 149, 112070, pp. 1–12.

Nyamnjoh, Francis B. (2017) 'Incompleteness: Frontier Africa and the Currency of Conviviality', *Journal of Asian and African Studies*, 52, 3, pp. 253–70.

Nyanzi, Stella (2014) 'Queering Queer Africa', in Zethu Matebeni (eds), *Reclaiming Afrikan: Queer Perspectives on Sexual and Gender Identities* (Athlone: Modjaji Books), pp. 61–5.

Nyawalo, Mich (2016) 'Afro-futurism and the Aesthetics of Hope in Bekolo's *Les Saignantes* and Kahiu's *Pumzi*', *Journal of the African Literature Association*, 10, 2, pp. 209–21.

Obadare, Ebenezer (2009) 'The Uses of Ridicule: Humour, "Infrapolitics" and Civil Society in Nigeria', *African Affairs*, 108, 431, pp. 241–61.

Obadare, Ebenezer (2010) 'State of Travesty: Jokes and the Logic of Socio-Cultural Improvisation in Africa', *Critical African Studies*, 2, 4, pp. 92–112.

O'Connell, Hugh Charles (2016) '"We Are Change": The Novum as Event in Nnedi Okorafor's *Lagoon*', *Cambridge Journal of Postcolonial Literary Inquiry*, 3, 3, pp. 291–312.

Odysseos, Louiza (2019) 'Stolen Life's Poetic Revolt', *Millennium*, 47, 3, pp. 341–72.

Ogude, James (2023) '"Aesthetics of the Earth": African Literature as a Witness to Post-colonial Ecology', in James Ogude and Tafadzwa Mushonga (eds), *Environmental Humanities of Extraction in Africa: Poetics and Politics of Exploitation* (Abingdon: Routledge), pp. 93–104.

Okereke, Chukwumerije (2021) *Climate Ambition beyond Emission Numbers, Taking Stock of Progress by Looking Inside Countries and Sectors, Nigeria* (Paris; Deep Decarbonization Pathways [DDP] Initiative-IDDRI, 2021).

Okereke, Chukwumerije (2024) 'Degrowth, Green Growth, and Climate Justice for Africa', *Review of International Studies*, 50, 5, pp. 910–20.

Okereke, Chukwumerije Alexia Coke, Mulu Geebreyesus, Tsegaye Ginbo, Jeremy J. Wakeford, and Yacob Mulugetta (2019) 'Governing Green Industrialisation in Africa: Assessing Key Parameters for a Sustainable Socio-technical Transition in the Context of Ethiopia', *World Development*, 115, pp. 279–90.

Okorafor, Nnedi (2009) 'Organic Fantasy', *African Identities*, 7, 2, pp. 275–86.

Okorafor, Nnedi (2011) 'Spider the Artist', *Lightspeed Magazine*, March. Available at http://www.lightspeedmagazine.com/fiction/spider-the-artist/ 27 June 2019.

Okorafor, Nnedi (2014) *Lagoon* (London: Hodder and Stoughton).

Okorafor, Nnedi (2015a) *The Book of Phoenix* (London: Hodder and Stoughton).

Okorafor, Nnedi (2015b) 'Writing Rage, Truth and Consequence', *Journal of the Fantastic in the Arts*, 26, 1, pp. 21–6.

Okorafor, Nnedi (2017) 'Sci-fi stories that Imagine a Future Africa', *TED Talk*, 22 November. Available at https://www.youtube.com/watch?v=Mt0PiXLvYlU (accessed 12 February 2020).

Okorafor, Nnedi (2018) *Who Fears Death* (London: HarperCollins). First published in 2010.

Okorafor, Nnedi (2019) *Africanfuturism Defined*, 19 October. Available at: http://nnedi.blogspot.com/2019/10/africanfuturism-defined.html (accessed 2 September 2020).

Okorafor, Nnedi (2021) *Noor* (New York: DAW Books).

Okri, Ben (2023) *Tiger Work: Stories, Essays and Poems about Climate Change* (London: Bloomsbury).

Oku, Arit (2021) 'Africanfuturism and the Reframing of Gender in the Fiction of Nnedi Okorafor', *Feminist Africa*, 2, 2, pp. 75–89.

Okungbowa, Suyi Davies (2019) 'Dune Song', *Apex Magazine*, 7 May. Available at https://apex-magazine.com/short-fiction/dune-song/ (accessed 3 August 2023).

Olaoluwa, Senayon (2018) 'The Human and the Non-human: African Sexuality Debate and Symbolisms of Transgression', in Zethu Matebeni, Surya Monro, and Vasu Reddy (eds), *Queer in Africa: LGBTQI Identities, Citizenship, and Activism* (Abingdon: Routledge), pp. 20–40.

Olaopa, Olawale R., Yusuf O. Akinwale, and Ibikunle O. Ogundari (2018) 'Governance Institutions for Sustainable Energy Resources Management in Nigeria: Issues, Perspectives and Policy Agenda for Action', in Lucky E. Asuelime and Andrew Emmanuel Okem (eds), *The Political Economy of Energy in Sub-Saharan Africa* (Abingdon: Routledge), pp. 82–101.

Omelsky, Matthew (2014) '"After the End Times": Postcrisis African Science Fiction', The *Cambridge Journal of Postcolonial Literary Inquiry*, 1, 1, pp. 33–49.

O'Neill, Brian C., Elmar Kriegler, Kristie L. Ebi, Eric Kemp-Benedict, Keywan Riahi, Dale S. Rothman, Bas J. van Ruijven, Detlef P. van Vuuren, Joern Birkmann, Kasper Kok, Marc Levy, and William Solecki (2017), 'The Roads Ahead: Narratives for Shared Socioeconomic

Pathways Describing World Futures in the 21st Century', *Global Environmental Change*, 42, pp. 169–80.

Onwualu, Chinelo (2019) 'What the Dead Man Said', *Slate*, 24 August. Available at https://slate.com/technology/2019/08/chinelo-onwualu-what-the-dead-man-said.html (accessed 25 September 2023).

Onwualu, Chinelo (2022) 'Letters to My Mother', in Francesco Verso (ed.), *Meteotopia: Futures of Climate (In)Justice* (Roma: Associazione Future Fiction), pp. 65–79.

Onyebuchi, Tochi (2019) *War Girls* (London: Razorbill).

Opondo, Sam Okoth (2016) 'Childhood, Redemption and the Prosaics of Waiting', in Anna M. Agathangelou and Kyle D. Killian (eds), *Time, Temporality and Violence in International Relations: (De)Fatalizing the Present, Forging Radical Alternatives* (Abingdon: Routledge), pp. 205–20.

Opondo, Sam Okoth (2022) '"I Was Worlds": Meditations on Diplomatic Becomings and Dissensual Friendship', *Globalizations*, pp. 1–21. Available at https://www.tandfonline.com/doi/full/10.1080/14747731.2022.2156239 (accessed 8 March 2024).

Oreskes, Naomi and Erik M. Conway (2014) *The Collapse of Western Civilisation: A View from the Future* (New York: Columbia University Press).

Osunmuyiwa, Olufolahan, Frank Biermann, and Agni Kalfagianni (2018) 'Applying the Multi-level Perspective on Socio-technical Transitions to Rentier States: The Case of Renewable Energy Transitions in Nigeria', *Journal of Environmental Policy & Planning*, 20, 2, pp. 143–56.

Osunmuyiwa, Olufolahan and Agni Kalfagianni (2017) 'The Oil Climax: Can Nigeria's Fuel Subsidy Reforms Propel Energy Transitions?', *Energy Research & Social Science*, 27, pp. 96–105.

Otto, Danny, Therese Thoni, Felix Wittstock, and Silke Beck (2021) 'Exploring Narratives on Negative Emissions Technologies in the Post-Paris Era', *Frontiers in Climate*, 3, 684135, pp. 1–13.

Otu, Kwame Edwin (2021) 'When the Lagoons Remember: An Afroqueer Futurist Reading of "Blue Ecologies of Agitation"', *Feminist Africa*, 2, 2, pp. 29–46.

Pahl, Miriam (2018) 'Time, Progress, and Multidirectionality in Nnedi Okorafor's *Who Fears Death*', *Research in African Literatures*, 49, 3, pp. 207–22.

Pak, Chris (2016) *Terraforming: Ecopolitical Transformations and Environmentalism in Science Fiction* (Liverpool: Liverpool University Press).

Patel, Raj and Jason W. Moore (2017) *A History of the World in Seven Cheap Things: A Guide to Capitalism, Nature, and the Future of the Planet* (Oakland: University of California Press).

Paterson, Matthew (2016) 'The Sociological Imagination of Climate Futures', in Paul Wapner and Hilal Elver (eds), *Reimagining Climate Change* (Abingdon: Routledge), pp. 14–28.

Paterson, Matthew (2021a) 'Climate Change and International Political Economy: Between Collapse and Transformation', *Review of International Political Economy*, 28, 2, pp. 394–405.

Paterson, Matthew (2021b) '"The End of the Fossil Fuel Age"? Discourse Politics and Climate Change Political Economy', *New Political Economy*, 26, 6, pp. 923–36.

Paterson, Matthew, Paul Tobin, and Stacy D. VanDeveer (2022) 'Climate Governance Antagonisms: Policy Stability and Repoliticization', *Global Environmental Politics*, 22, 2, pp. 1–11.

Pathak, Siddharth (2017) 'Why Develop 2050 Pathways?' *UNFCCC 2050 Pathways Platform*, 8 July.

Pathak, Sneha (2013) 'The Road Less Travelled: *Mara and Dann* as a Heroine's Quest Narrative', *European Academic Research*, 1, 8, pp. 2222–31.

Pattee, Emma (2023) 'On the False Promise of Climate Fiction', *Literary Hub*, 14 November. Available at lithub.com https://lithub.com/on-the-false-promise-of-climate-fiction/ (accessed 18 November 2023).

Pearse, Rebecca (2021) 'Theorising the Political Economy of Energy Transformations: Agency, Structure, Space, Process', *New Political Economy*, 26, 6, pp. 951–63.

Pereira, Laura, Tanja Hichert, Maike Hamann, Rika Preiser, and Reinette Biggs (2018) 'Using Futures Methods to Create Transformative Spaces: Visions of a Good Anthropocene in Southern Africa', *Ecology and Society*, 23, 1, pp. 1–13.

Pereira, Laura, Nadia Sitas, Federica Ravera, Amanda Jimenez-Aceituno, and Andrew Merrie (2019) 'Building Capacities for Transformative Change towards Sustainability: Imagination in Intergovernmental Science-Policy Scenario Processes', *Elementa: Science of the Anthropocene*, 7, 35, pp. 1–19.

Pereira, Laura, Charne Lavery, Busiso Moyo, Odirilwe Selomane, Nadia Sitas, Rike Sitas, and Christopher Trisos (2021) 'Wakanda Phambili!: African Science Fiction for Reimagining the Anthropocene', in Sandra Kemp and Jenny Andersson (eds), *Futures* (Oxford: Oxford University Press), pp. 262–80.

Peterson, V. Spike (2020) 'Family Matters in Racial Logics: Tracing Intimacies, Inequalities, and Ideologies', *Review of International Studies*, 46, 2, pp. 177–96.

Pheko, Lebohang Liepollo (2024) 'The Political Economy of Land Reparations in South Africa', *Development and Change*, 55, 4, pp. 800–29.

Pierre-Louis, Kendra (2020) '"Wakanda Doesn't Have Suburbs": How Movies Like *Black Panther* Could Help Us Save the Planet', *Time* (US), 21 September. Available at https://time.com/5889324/movies-climate-change/ (accessed 8 March 2024).

Pieterse, Edgar (2011) 'Grasping the Unknowable: Coming to Grips with African Urbanisms', *Social Dynamics*, 37, 1, pp. 5–23.

Pin-Fat, Véronique (2016) 'Dissolutions of the Self', Naeem Inayatullah and Elizabeth Dauphinee (eds) *Narrative Global Politics: Theory, History and the Personal in International Relations* (London: Routledge), pp. 25–34.

Plumwood, Val (2004) 'Taken by a Crocodile', *The Age*, 12 January. Available at https://www.theage.com.au/national/taken-by-a-crocodile-20040112-gdx34k.html (accessed 14 December 2023).

Power, Marcus, Peter Newell, Lucy Baker, Harriet Bulkeley, Joshua Kirshner, and Adrian Smith (2016) 'The Political Economy of Energy Transitions in Mozambique and South Africa: The Role of the Rising Powers', *Energy Research & Social Science*, 17, pp. 10–19.

Pratchett, Terry and Neil Gaiman (2014) *Good Omens* (London: Transworld Publishers).

Primorac, Ranka (2016) 'In Pursuit of Transcendence: Honouring Doris Lessing—Introduction', *Journal of Southern African Studies*, 42, 1, pp. 109–10.

Propst, Lisa (2017) 'Information Glut and Conspicuous Silence in Lauren Beukes's *Zoo City*', *Journal of Postcolonial Writing*, 53, 4, pp. 414–26.

Pulver, Simone and Stacy D. VanDeveer (2009) '"Thinking about Tomorrows": Scenarios, Global Environmental Politics, and Social Science Scholarship', *Global Environmental Politics*, 9, 2, pp. 1–13.

Punter, David (2016) 'Lessing and Time Travel', in Kevin Brazil, David Sergeant, and Tom Sperlinger (eds), *Doris Lessing and the Forming of History* (Edinburgh: Edinburgh University Press), pp. 128–36.

Rajaniemi, Hannu (2017) 'Making Stranger Worlds', *Nature*, 552, p. 332.

Ramutsindela, Maano (2007) *Transfrontier Conservation in Africa: At the Confluence of Capital, Politics and Nature* (Cambridge, MA: CABI).

Rao, Rahul (2014) 'Queer Questions', *International Feminist Journal of Politics*, 16, 2, pp. 199–217.

Rao, Rahul (2020) *Out of Time: The Queer Politics of Postcoloniality* (Oxford: Oxford University Press).

Ravecca, Paulo and Elizabeth Dauphinee (2018) 'Narrative and the Possibilities for Scholarship', *International Political Sociology*, 12, pp. 125–38.

Raven, Paul Graham (2017) 'Telling Tomorrows: Science Fiction as an Energy Futures Research Tool', *Energy Research & Social Science*, 31, pp. 164–9.

Raven, Paul Graham and Shirin Elahi (2015) 'The New Narrative: Applying Narratology to the Shaping of Futures Output', *Futures*, 74, pp. 49–61.

Raven, Paul and Johannes Stripple (2021) 'Touring the Carbon Ruins: Towards an Ethics of Speculative Decarbonisation', *Global Discourse*, 11, 1–2, pp. 221–40.

Reddy, Vasu, Surya Monro, and Zethu Matebeni (2018) 'Introduction', in Zethu Matebeni, Surya Monro, and Vasu Reddy (eds), *Queer in Africa: LGBTQI Identities, Citizenship, and Activism* (Abingdon: Routledge), pp. 1–16.

Redifer, Lindsay (2016) 'Standing Still', in Manjana Milkoreit, Meredith Martinez, and Joey Eshrich (eds), *Everything Change: An Anthology of Climate Fiction* (Phoenix: Arizona State University), pp. 179–96

Rennkamp, Britta (2019) 'Power, Coalitions and Institutional Change in South African Climate Policy', *Climate Policy*, 19, 6, pp. 756–70.

Republic of South Africa (2020) *South Africa's Low Emission Development Strategy*, February. Available at https://unfccc.int/sites/default/files/resource/South%20Africa%27s%20Low%20Emission%20Development%20Strategy.pdf (accessed 24 June 2022).

Resnick, Mike (1998) *Kirinyaga: A Fable of Utopia* (New York: Ballentine).

Rico, Amanda Renée (2017) 'Gendered Ecologies and Black Feminist Futures in Wanuri Kahiu's *Pumzi*, Wangechi Mutu's "The End of Eating Everything", and Ibi Zoboi's "The Farming of Gods"', *Wagadu: A Journal of Transnational Women's and Gender Studies*, 18, pp. 81–99.

Rizk, Yara (2023) 'COP28: Loss and Damage Fund both "Symbolic and Useful" for Africa', *The Africa Report*, 19 December. Available at https://www.theafricareport.com/330505/cop28-loss-and-damage-fund-both-symbolic-and-useful-for-africa/ (accessed 24 January 2024).

Robinson, Blake and Karl Schulschenk (2017) 'African Alternatives', *postfossil.city competition*. Available at https://postfossil.city/en/finalists/african-alternatives (accessed 29 September 2023).

Robinson, Kim Stanley (1992) *Red Mars* (London: Random House).

Robinson, Kim Stanley (1993) *Green Mars* (London: Random House).

Robinson, Kim Stanley (1996) *Blue Mars* (London: Random House).

Robinson, Kim Stanley (2017) *New York 2140* (London: Orbit).

Robinson, Kim Stanley (2020) *Ministry of the Future* (London: Orbit).

Robinson, Kim Stanley (2019) 'Climate Change Forces Post-capitalism', in Kum-Kum Bhavnani, John Foran, Priya A. Kurian, and Debashish Munshi (eds), *Climate Futures: Re-imagining Global Climate Justice* (London: Zed), pp. 46–53.

Rooney, Darrell (1998) *The Lion King II: Simba's Pride* [film] (Walt Disney films).

Ross, Katherine and Taryn Fransen (2017) *Early Insights on Long-Term Climate Strategies*, Working Paper (Washington, DC: World Resources Institute).

Ross, Katie, Clea Schumer, Taryn Fransen, and Cynthia Elliott (2021) '10 Big Findings from the First Batch of Long-Term Strategies', *World Resources Institute*, 20 September. Available at https://www.wri.org/insights/findings-long-term-strategies-first-batch (accessed 19 May 2023).

Rothe, Delf (2020) 'Governing the End Times? Planet Politics and the Secular Eschatology of the Anthropocene', *Millennium*, 48, 2, pp. 143–64.

Royo, Margarita Garfias, Loan Diep, Joe Mulligan, Pascal Mukanga, and Priti Parikh (2022) 'Linking the UN Sustainable Development Goals and African Agenda 2063: Understanding Overlaps and Gaps between the Global Goals and Continental Priorities for Africa', *World Development Sustainability*, 1, 100010, pp. 1–9.

Rutazibwa, Olivia U. and Robbie Shilliam (2018) 'Postcolonial Politics: An Introduction', in Olivia U. Rutazibwa and Robbie Shilliam (eds), *Routledge Handbook of Postcolonial Politics* (Abingdon: Routledge), pp. 1–15.

Sabaratnam, Meera (2011) 'IR in Dialogue ... but Can We Change the Subjects? A Typology of Decolonising Strategies for the Study of World Politics', *Millennium*, 39, 3, pp. 781–803.

Sabaratnam, Meera and Mark Laffey (2023) 'Complex Indebtedness: Justice and the Crisis of Liberal Order', *International Affairs*, 99, 1, pp. 161–80.

Said, Edward (1994) *Culture and Imperialism* (London: Vintage).

Sakakibara, Chie (2023) 'All Things Sacred: Love, Resilience, and Sovereignty in Linda Infante Lyons's Alaska Native Icon Series', *Journal of Indigenous Social Development*, 12, 1, pp. 28–46.

Salami, Minna (2021) 'The Liquid Space where African Feminism and African Futurism Meet', *Feminist Africa*, 2, 2, pp. 90–4.

Salmon, Christian (2010) *Storytelling: Bewitching the Modern Mind*, translated by David Macey (London: Verso).

Samatar, Sofia (2017) 'Toward a Planetary History of Afrofuturism', *Research in African Literatures*, 48, 4, pp. 175–91.

Sandilands, Catriona (2017) 'Feminism and Biopolitics: A Cyborg Account', in Sherilyn MacGregor (ed.), *Routledge Handbook of Gender and Environment* (Abingdon: Routledge), pp. 229–38.

Sanneh, Alieu B. (2022) 'Election, Ethnic Voting and Regime Change in The Gambia', *Third World Quarterly*, 43, 12, pp. 2775–90.

Sargeant, David (2016) 'Lessing and the Scale of Environmental Crisis', in Kevin Brazil, David Sergeant, and Tom Sperlinger (eds), *Doris Lessing and the Forming of History* (Edinburgh: Edinburgh University Press), pp. 111–27.

Satterthwaite, David (2016) 'Will Africa Have the World's Largest Cities in 2100?', *IIED insight*. Available at https://www.iied.org/will-africa-have-worlds-largest-cities-2100 (accessed 13 September 2024).

Schneider-Mayerson, Matthew (2018) 'The Influence of Climate Fiction: An Empirical Survey of Readers', *Environmental Humanities*, 10, 2, pp. 473–500.

Schneider-Mayerson, Matthew (2019) 'Whose Odds? The Absence of Climate Justice in American Climate Fiction Novels', *ISLE: Interdisciplinary Studies in Literature and Environment*, 26, 4, pp. 944–67.

SEI, Climate Analytics, E3G, IISD and UNEP (2023) *The Production Gap: Phasing down or Phasing up? Top Fossil Fuel Producers Plan Even More Extraction Despite Climate Promises* (Stockholm: Stockholm Environment Institute, Climate Analytics, E3G, International Institute for Sustainable Development and United Nations Environment Programme).

Selbin, Eric (2019) 'Resistance and Revolution in the Age of Authoritarian Revanchism: The Power of Revolutionary Imaginaries in the Austerity-Security State Era', *Millennium*, 47, 3, pp. 483–96.

Seymour, Nicole (2013) *Strange Natures: Futurity, Empathy, and the Queer Ecological Imagination* (Chicago: University of Illinois Press).

Seymour, Nicole (2017) 'Transgender Environments', in Sherilyn MacGregor (ed.), *Routledge Handbook of Gender and Environment* (Abingdon: Routledge), pp. 253–69.

Seymour, Nicole (2018) *Bad Environmentalism: Irony and Irreverence in the Ecological Age* (Minneapolis: University of Minnesota Press).

Sexton, Jared (2016) 'The Social Life of Social Death: On Afro-pessimism and Black Optimism', in Anna M. Agathangelou and Kyle D. Killian (eds), *Time, Temporality and Violence in International Relations: (De)Fatalizing the Present, Forging Radical Alternatives* (Abingdon: Routledge), pp. 61–75.

Shapiro, Michael J. (2010) *The Time of the City: Politics, Philosophy and Genre* (Abingdon: Routledge).

Shirley, Rebekah, Yared Deme, and Susan Onyango (2023) 'Key COP28 Takeaways for Africa', *World Resources Institute*, 20 December. Available at https://www.wri.org/africa/key-cop28-takeaways-africa (accessed 24 January 2024).

Sides, Kirk Bryan (2019) 'Seed Bags and Storytelling: Modes of Living and Writing after the End in Wanuri Kahiu's *Pumzi*', *Critical Philosophy of Race*, 7, 1, pp. 107–23.

Simmet, Hilton R. (2018) '"Lighting a Dark Continent": Imaginaries of Energy Transition in Senegal', *Energy Research & Social Science*, 40, pp. 71–81.

Simone, AbdouMaliq and Edgar Pieterse (2017) *New Urban Worlds: Inhabiting Dissonant Times* (Cambridge: Polity).

Simoneti, Caio A. Martins (2022) '"More Human than Human": Colonial Logics and the Modern Subject in Science Fiction Films', *Millennium*, 51, 1, pp. 184–211.

Sinthumule, Ndidzulafhi Innocent (2016) 'Local Community Disempowerment at the (Trans)frontier', in Maano Ramutsindela, Giorgio Miescher, and Melanie Boehi (eds), *The Politics of Nature and Science in Southern Africa* (Basle: Basler Afrika Bibliographien), pp. 231–51.

Slaymaker, William (2001) 'Ecoing the Other(s): The Call of Global Green and Black African Responses', *PMLA*, 116, 1, pp. 129–44.

Smith, Brady (2016) 'SF, Infrastructure, and the Anthropocene: Reading Moxyland and Zoo City', *Cambridge Journal of Postcolonial Literary Inquiry*, 3, 3, pp. 345–59.

Smith, Eric (2012) *Globalization, Utopia, and Postcolonial Science Fiction: New Maps of Hope* (New York: Palgrave Macmillan).

Sperlinger, Tom (2017) 'Radical Pedagogy in Doris Lessing's *Mara and Dann*', *Critique: Studies in Contemporary Fiction*, 58, 3, pp. 300–11.

Steele, Meili (2011) 'Literature as Public Reasoning in the Political Struggles over Imagination', in Chiara Bottici and Benoît Challand (eds), *The Politics of Imagination* (Abingdon: Birkbeck Law Press), pp. 178–94.

Stobie, Cheryl (2012) 'Dystopian Dreams from South Africa: Lauren Beukes's *Moxyland* and *Zoo City*', *African Identities*, 10, 4, pp. 367–80.

Stoetzler, Marcel and Nira Yuval-Davis (2002) 'Standpoint Theory, Situated Knowledge and the Situated Imagination', *Feminist Theory*, 3, 3, pp. 315–33.

Strange, Susan (1982) 'Cave! Hic Dragones: A Critique of Regime Analysis', *International Organization*, 36, 2, pp. 479–96.

Strausz, Erzsébet (2024) 'Transforming Epistemological Disconnection from the More-than-Human World: (Inter)nodes of Ecologically Attuned Ways of Knowing', *International Relations*, 38, 3, pp. 407–26.

Streeby, Shelley (2018) *Imagining the Future of Climate Change: World-making through Science Fiction and Activism* (Oakland: University of California Press).

Sturgeon, Noël (2004) '"The Power Is Yours, Planeteers!" Race, Gender and Sexuality in Children's Environmentalist Popular Culture', in Rachel Stein (ed.), *New Perspectives on Environmental Justice: Gender, Sexuality, and Activism* (New Brunswick: Rutgers University Press), pp. 262–76.

Sultana, Farhana (2022) 'The Unbearable Heaviness of Climate Coloniality', *Political Geography*, 99, 102638, pp. 1–16.

Suvin, Darko (2016) *Metamorphoses of Science Fiction: On the Poetics and History of a Literary Genre*, edited by Gerry Canavan (Bern: Peter Lang).

Svoboda, Michael (2016) 'Cli-fi on the Screen(S): Patterns in the Representations of Climate Change in Fictional Films', *WIREs Climate Change*, 7, pp. 43–64.

Swilling, Mark and Eve Annecke (2012) *Just Transitions: Explorations of Sustainability in an Unfair World* (Claremont: UCT Press).

Swilling, Mark, Josephine Musango, and Jeremy Wakeford (2016) 'Developmental States and Sustainability Transitions: Prospects of a Just Transition in South Africa', *Journal of Environmental Policy & Planning*, 18, 5, pp. 650–72.

Sylla, Ndongo Samba, Andrew M. Fischer, Annina Kaltenbrunner, and Sreerekha Sathi (2024) 'Global Reparations within Capitalism: Aspirations and Tensions in Contemporary Movements for Reparatory Justice', *Development and Change*, 55, 4, pp. 560–600.

Sze, Julie (2020) *Environmental Justice in a Moment of Danger* (Oakland: University of California Press).

Táíwò, Olúfẹmi O. (2022) *Reconsidering Reparations* (Oxford: Oxford University Press).

Tamale, Sylvia (2020) *Decolonization and Afro-Feminism* (Ottawa: Daraja Press).

Tanpepper, Saul (2021) 'The Cloud Weaver's Song', *Imagine 2200*, 14 September. Available at https://grist.org/fix/arts-culture/imagine-2200-climate-fiction-cloud-weavers-song/ (accessed 24 January 2024).

Tayeb, Lamia (2009) 'Arabian Nights Fairy-tale Turned Postcolonial Parable: Narrative Manoeuvres in Doris Lessing's "Mara and Dann"', *Doris Lessing Studies*, 28, 2, pp. 18–25.

Taylor, Charles (2002) 'Modern Social Imaginaries', *Public Culture*, 14, 1, pp. 91–124.

Thaler, Mathias (2022) *No Other Planet: Utopian Visions for a Climate-Changed World* (Cambridge: Cambridge University Press).

The Economist (2020) 'Africa's population will double by 2050, 26 March. Available at https://www.economist.com/special-report/2020/03/26/africas-population-will-double-by-2050 (accessed 28 November 2024).

Thompson, Kim-Ly and Natalie C. Ban (2022) '"Turning to the Territory": A Gitga'at Nation Case Study of Indigenous Climate Imaginaries and Actions', *Geoforum*, 137, pp. 230–6.

Thompson, Tade (2016) *Rosewater* (Lexington: Apex Books).

Thompson, Tade (2019a) *The Rosewater Insurrection* (London: Orbit Books).

Thompson, Tade (2019b) *The Rosewater Redemption* (London: Orbit Books).

Tilly, Charles (2002) *Stories, Identities, and Political Change* (Lanham: Rowman & Littlefield Publishers).

Tobin, Paul, Nicole M. Schmidt, Jale Tosun, and Charlotte Burns (2018) 'Mapping States' Paris Climate Pledges: Analysing Targets and Groups at COP 21', *Global Environmental Change*, 48, pp. 11–21.

Tozer, Laura and Nicole Klenk (2018) 'Discourses of Carbon Neutrality and Imaginaries of Urban Futures', *Energy Research & Social Science*, 35, pp. 174–81.

Trexler, Adam (2015) *Anthropocene Fictions: The Novel in a Time of Climate Change* (Charlottesville: University of Virginia Press).

Trisos, Christopher H. et al. (2022) 'Africa', in IPCC, *Climate Change 2022: Impacts, Adaptation and Vulnerability*, Contribution of Working Group II to the Sixth Assessment Report of the Intergovernmental Panel on Climate Change (Cambridge: Cambridge University Press), pp. 1285–455.

Trollip, Hilton (2021) *Climate Ambition beyond Emission Numbers, Taking Stock of Progress by Looking Inside Countries and Sectors, South Africa* (Paris: Deep Decarbonization Pathways [DDP] Initiative-IDDRI).

Tsamaase, Tlotlo (2016) 'Virtual Snapshots', *Vice*, 18 February. Available at https://www.vice.com/en/article/aekvne/virtual-snapshots (accessed 5 May 2021).

Tsamaase, Tlotlo (2019) 'Eclipse Our Sins', *Clarkesworld*, 159, December. Available at https://clarkesworldmagazine.com/tsamaase_12_19/ (accessed 2 August 2023).

Tsamaase, Tlotlo (2022) 'Eco-humans', in Francesco Verso (ed.), *Meteotopia: Futures of Climate (In)Justice* (Roma: Associazione Future Fiction), pp. 15–28.

Tschakert, Petra (2022) 'More-than-Human Solidarity and Multispecies Justice in the Climate Crisis', *Environmental Politics*, 31, 2, pp. 277–96.

Tuck, Eve and K. Wayne Yang (2012) 'Decolonization Is Not a Metaphor', *Decolonization: Indigeneity, Education & Society*, 1, 1, pp. 1–40.

Tyler, Emily and Kathryn Hochstetler (2021) 'Institutionalising Decarbonisation in South Africa: Navigating Climate Mitigation and Socio-economic Transformation', *Environmental Politics*, 30, 1, pp. 184–205.

UNDP (2017) *Africa and Foresight: Better Futures in Development* (Singapore: GCPSE).

UNFCCC (2023) 'Long-term Strategies Portal.' Available at https://unfccc.int/process/the-paris-agreement/long-term-strategies (accessed 12 May 2023).

Urry, John (2016) *What Is the Future?* (Cambridge: Polity Press).

van der Helm, Ruud (2009) 'The Vision Phenomenon: Towards a Theoretical Underpinning of Visions of the Future and the Process of Envisioning', *Futures*, 41, pp. 96–104.

van Noorloos, Femke and Marjan Kloosterboer (2018) 'Africa's New Cities: The Contested Future of Urbanisation', *Urban Studies*, 55, 6, pp. 1223–41.

van Ryneveld, Tara Nair, and Mine Islar (2022) 'Coloniality as a Barrier to Climate Action: Hierarchies of Power in a Coal-Based Economy', *Antipode*, 55, 3, pp. 958–81.

van Veuren, Mocke Jansen (2012) 'Tooth and Nail: Anxious Bodies in Neill Blomkamp's *District 9*', *Critical Arts: South-North Cultural and Media Studies*, 26, 4, pp. 570–86.

Van de Graaf, Thijs, (2022) 'Transition Troubles: Petrostates, Decarbonisation, and the Geopolitics of Peak Oil Demand', in Roland Dannreuther and Wojciech Ostrowski (eds), *Handbook on Oil and International Relations* (Cheltenham: Edward Elgar), pp. 427–44.

VanderMeer, Jeff (2023) 'Climate Fiction Won't Save Us', *Esquire*, 19 April. Available at https://www.esquire.com/entertainment/books/a43541988/climate-fiction-wont-save-us/ (accessed 18 November 2023).

Veracini, Lorenzo (2011) '*District 9* and *Avatar*: Science Fiction and Settler Colonialism', *Journal of Intercultural Studies*, 32, 4, pp. 355–67.

Vergès, Françoise (2002) 'The Power of Words', *Public Culture*, 14, 3, pp. 607–10.

Vervoort, Joost M., Roy Bendor, Aisling Kelliher, Oscar Strik, and Ariella E.R. Helfgott (2015) 'Scenarios and the Art of Worldmaking', *Futures*, 74, pp. 62–70.

Vervoort, Joost and Aarti Gupta (2018) 'Anticipating Climate Futures in a 1.5°C Era: The Link between Foresight and Governance', *Current Opinion in Environmental Sustainability*, 31, pp. 104–11.

Wa Thiong'o, Ngũgĩ (1977) *Petals of Blood* (London: Heinemann).

Walder, Dennis (2008) '"Alone in a Landscape": Lessing's African Stories Remembered', *Journal of Commonwealth Literature*, 43, 2, pp. 99–115.

Wallace-Wells, David (2019) *The Uninhabitable Earth: A Story of the Future* (London: Allen Lane).

Watkins, Susan (2006) '"Grande Dame" or "New Woman": Doris Lessing and the Palimpsest', *Literature Interpretation Theory*, 17, 3–4, pp. 243–62.

Watkins, Susan (2010) *Doris Lessing* (Manchester: Manchester University Press).

Watkins, Susan (2012) 'Future Shock: Rewriting the Apocalypse in Contemporary Women's Fiction', *Literature Interpretation Theory*, 23, 2, pp. 119–37.

Watson, Vanessa (2013) 'African Urban Fantasies: Dreams or Nightmares?', *Environment & Urbanization*, 26, 1, pp. 215–31.

Weheliye, Alexander G. (2014) *Habeas Viscus: Racializing Assemblages, Biopolitics, and Black Feminist Theories of the Human* (Durham: Duke University Press).

Weik von Mossner, Alexa (2017) *Affective Ecologies: Empathy, Emotion, and Environmental Narrative* (Columbus: Ohio State University Press).

Weldes, Jutta (1999) 'Going Cultural: *Star Trek*, State Action, and Popular Culture', *Millennium*, 28, 1, pp. 117–34.

Weldes, Jutta (2001) 'Globalisation Is Science Fiction', *Millennium*, 30, 3, pp. 647–67.

Weldes, Jutta and Christina Rowley (2015) 'So, How Does Popular Culture Relate to World Politics?', in Federica Caso and Caitlin Hamilton (eds), *Popular Culture and World Politics: Theories, Methods, Pedagogies* (Bristol: E-IR), pp. 11–34.

Welland, Julia (2018) 'Joy and War: Reading Pleasure in Wartime Experiences', *Review of International Studies*, 44, 3, pp. 438–55.

Wenzel, Jennifer (2006) 'Petro-Magic-Realism: Toward a Political Ecology of Nigerian Literature', *Postcolonial Studies*, 9, 4, pp. 449–64.

White, Hayden (2011) 'Modern Politics and the Historical Imaginary', in Chiara Bottici and Benoît Challand (eds), *The Politics of Imagination* (Abingdon: Birkbeck Law Press,), pp. 162–77.

Whyte, Kyle P. (2018) 'Indigenous Science (Fiction) for the Anthropocene: Ancestral Dystopias and Fantasies of Climate Change Crises', *Environment and Planning E: Nature and Space*, 1, 1–2, pp. 224–42.

Whyte, Kyle P. (2020) 'Too Late for Indigenous Climate Justice: Ecological and Relational Tipping Points', *WIREs Climate Change*, wcc.603, pp. 1–7.

Williams, Jim and Henri Waisman (2017) *2050 Pathways: A Handbook, UNFCCC 2050 Pathways Platform*, 8 July. Available at https://www.2050pathways.org/resources/2050-pathways-handbook/ (accessed 7 March 2019).

Williams, Saul and Anisia Uzeyman (2021) *Neptune Frost* [film] (Kino Lorber).

Wilson, Jessica (2019) 'The Drought', in Karen O'Brien, Ann El Khoury, Nicole Schafenacker, and Jordan Rosenfeld (eds), *Our Entangled Future: Stories to Empower Quantum Social Change* (Oslo: AdaptationCONNECTS Research Project, University of Oslo), pp. 31–40.

Wilson-Tagoe, Nana (1997) 'Reading towards a Theorization of African Women's Writing: African Women Writers within Feminist Gynocriticism', in Stephanie Newell (ed.), *Writing African Women: Gender, Popular Culture and Literature in West Africa* (London: Zed Books), pp. 11–28.

Winkler, Harald (2022) 'South Africa Could Produce a Lot More Renewable Energy: Here's What It Needs', *The Conversation*, 12 July. Available at https://theconversation.com/south-africa-could-produce-a-lot-more-renewable-energy-heres-what-it-needs-185897 (accessed 5 October 2022).

Womack, Ytasha (2013) *Afrofuturism: The World of Black Sci-Fi and Fantasy Culture* (Chicago: Lawrence Hill Books).

Wood, Nick and Faeza Meyer (2022) 'Just Stories: The Role of Speculative Fiction in Challenging the Growing Climate Apartheid', *Psychology in Society*, 63, pp. 29–51.

Wood, Tilly Walker, Katharina Richter, and Ed Atkins (2024) 'Modelling beyond Growth Perspectives for Sustainable Climate Futures: The Case for Rethinking Shared Socioeconomic Pathways', *Energy Research & Social Science*, 117, pp. 1–5.

Wright, Christopher, Daniel Nyberg, Christian de Cock, and Gail Whiteman (2013) 'Future Imaginings: Organizing in Response to Climate Change', *Organization*, 20, 5, pp. 647–58.

Wright, Laura (2010) *Wilderness into Civilised Shapes: Reading the Postcolonial Environment* (Athens: University of Georgia Press).

Wright, Laura (2011) 'Inventing Tradition and Colonizing the Plants: Ngugi wa Thiong'o's *Petals of Blood* and Zakes Mda's *The Heart of Redness*', in Byron Caminero-Santangelo and Garth Myers (eds), *Environment at the Margins: Literary and Environmental Studies in Africa* (Athens, OH: Ohio University Press), pp. 235–56.

Wynter, Sylvia (1987) 'On Disenchanting Discourse: "Minority" Literary Criticism and Beyond', *Cultural Critique*, 7, pp. 207–44.

Wynter, Sylvia (2003) 'Unsettling the Coloniality of Being/Power/Truth/Freedom: Towards the Human, after Man, Its Overrepresentation—An Argument', *CR: The New Centennial Review*, 3, 3, pp. 257–337.

Wynter, Sylvia and Katherine McKittrick (2015) 'Unparalleled Catastrophe for Our Species? Or, to Give Humanness a Different Future: Conversations', in Katherine McKittrick (ed.), *Sylvia Wynter: On Being Human as Praxis* (Durham: Duke University Press), pp. 9–89.

Yoder, Kate (2023) 'The Summer that Reality Caught up with Climate Fiction', *Grist*, 22 September. Available at https://grist.org/culture/summer-reality-caught-climate-fiction-heatwave/ (accessed 8 March 2024).

Youatt, Rafi (2014) 'Interspecies Relations, International Relations: Rethinking Anthropocentric Politics', *Millennium*, 43, 1, pp. 207–23.

Young, Iris Marion (1984) 'Pregnant Embodiment: Subjectivity and Alienation', *The Journal of Medicine and Philosophy*, 9, pp. 45–62.

Young, Kevin L. and Charli Carpenter (2018) 'Does Science Fiction Affect Political Fact? Yes and No: A Survey Experiment on "Killer Robots"', *International Studies Quarterly*, 62, pp. 562–76.

Yusoff, Kathryn (2010) 'Biopolitical Economies and the Political Aesthetics of Climate Change', *Theory, Culture and Society*, 27, 2–3, pp. 73–99.

Yusoff, Kathryn (2018) *A Billion Black Anthropocenes or None* (Minneapolis: University of Minnesota Press).

Yusoff, Kathryn and Jennifer Gabrys (2011) 'Climate Change and the Imagination', *WIREs Climate Change*, 2, 4, pp. 516–34.

Zamalin, Alex (2019) *Black Utopia: The History of an Idea from Black Nationalism to Afrofuturism* (New York: Columbia University Press).

Zeleza, Paul Tiyambe (2018) '*Black Panther* and the Persistence of the Colonial Gaze', *The Medium*, 3 April. Available at https://medium.com/@USIUAfrica/black-panther-and-the-persistence-of-the-colonial-gaze-6c093fa4156d (accessed 18 May 2020).

Index

For the benefit of digital users, indexed terms that span two pages (e.g., 52–53) may, on occasion, appear on only one of those pages.